The Construction (Design and Management) Regulations 1994 Explained

Second edition

Raymond Joyce BSc, MSc, LLB, DIC, CEng, FICE, ACIArb
Chartered civil engineer and solicitor of the Supreme Court

 ThomasTelford

Published by Thomas Telford Publishing, Thomas Telford Ltd, 1 Heron Quay, London E14 4JD.

URL: http://www.thomastelford.com

Distributors for Thomas Telford books are
USA: ASCE Press, 1801 Alexander Bell Drive, Reston, VA 20191-4400, USA
Japan: Maruzen Co. Ltd, Book Department, 3–10 Nihonbashi 2-chome, Chuo-ku, Tokyo 103
Australia: DA Books and Journals, 648 Whitehorse Road, Mitcham 3132, Victoria

First edition 1995; reprinted 1995, 1997
Second edition 2001; reprinted 2003, 2005

A catalogue record for this book is available from the British Library

ISBN: 0 7277 3036 3

Typeset by Alex Lazarou, Surbiton, Surrey
Printed and bound in Great Britain by MPG Books, Bodmin, Cornwall

In my judgment the most striking feature is the clear lack of liaison within and between the defendants in relation to matters of safety. This is in no way inconsistent with the existence of an undoubtedly close working relationship...

'The Abbeystead Disaster'
Eckersley & Others -v- Binnie & Others
Rose J

Foreword

This book is an interesting, practical and comprehensive guide to the latest and most important development in health and safety law covering the construction industry – The Construction (Design and Management) Regulations 1994 – which came into force on 31st March 1995. The author, Raymond Joyce, demonstrates his ability to interpret and communicate complex issues in straightforward terms, which reflects his knowledge, experience and understanding both of construction practice and of the law.

The construction industry has always had the highest incident rate in respect of fatal accidents and serious injuries of all industries, with perhaps the exception of deep sea fishing. Since the beginning of this century H.M. Factory Inspectorate, and subsequently the Health and Safety Executive, have analysed reports of accidents from which they have identified the various causations. Legislation has been developed from such studies which, if adhered to, would substantially reduce or even eliminate many types of accident. Whilst the 1961 and 1966 Construction Regulations catered well for current main causes of accidents, particularly when supported by the Health and Safety at Work etc. Act 1974, the advancement of technology, the development of sophisticated plant, new construction techniques, the increased size and complexity of construction works and the improvements in the recognition of risks and hazards, meant that there was still scope for improvements in the industry's safety record.

Both legislators and safety professionals in the industry have recognised for some time that many gaps could be plugged and the accident record improved if obligations were placed on those indirectly involved in construction as well as those actually carrying out the works. Designers, architects and particularly clients influence the construction process. If that influence were used with accident prevention in mind, not only during the construction phase but throughout the life of the works until the demise with the demolition phase, a great contribution would have been made to the avoidance of accidents. The Federation of Demolition Contractors has been advocating for over 20 years the establishment of an as-built record file and will be delighted that this has now come to fruition.

The Construction (Design and Management) Regulations 1994 fill those gaps. I have no doubt that they will work and that we shall see a substantial improvement in the accident record of the construction industry. This book is essential reading for anybody concerned with the construction industry. I found it easy to understand and absorb and would particularly recommend it to those who until now may have been on the fringe of construction health and safety, such as architects, those concerned with building, civil engineering and structural design, building surveyors, quantity surveyors and construction equipment designers. Further, it is a 'must' for those responsible for the procurement or management of construction projects. The requirements and duties of the various parties are clearly explained by Raymond Joyce and supported by useful checklists.

A comparison may be made between these Regulations and the health and safety management during the construction of the Channel Tunnel. The accident record on the Channel Tunnel project witnessed a much lower incident rate than in the construction industry generally, due, I am sure, to the method of operation, particularly the co-operation between all parties in health and safety matters. A health and safety plan was devised and continually updated, approved by the safety authority and accepted by the client before execution. The safety authority on that project assumed a role similar in part to that of the planning supervisor.

Major civil engineering consultants and architectural practices will no doubt compete for this position and this book would be a most useful asset to their libraries. It is essential reading, too, for all students whose ambition it is to climb the management ladder of the construction industry. For many years I have campaigned for the inclusion of health and safety matters in first degree civil and structural engineering courses to breed a professional attitude to the topic. I would recommend this book as a useful addition to relevant course reading lists.

Kenneth Tomasin

Former Head of the Construction Engineering Branch of the Health and Safety Executive

Preface to the second edition

The debate surrounding the impact and efficiency, or otherwise, of the Construction (Design and Management) Regulations 1994 (referred to throughout the book as 'the Regulations') has continued almost without any interruption since their introduction on 31 March 1995. This has in part been due to the grave public concerns over safety standards within the industry generally and the government's campaign for improvements in the health and safety record of the construction industry. More cynically, the construction industry has pressed for a review of the Regulations because they see the impact on rising costs and a growing administrative burden. There are still many in the construction industry who refuse to believe, in my opinion incorrectly, that the Regulations have made any difference.

Since the first edition of this book, the role of designer has been reviewed by the Court of Appeal thus exposing a loophole that was quickly plugged by the Construction (Design and Management) (Amendment) Regulations 2000. There have also been a number of amendments to other related health and safety Regulations, all of which have led to consequential amendments to the Regulations. There have also been hundreds of prosecutions, which reveal something of the application of the Regulations in practice and provide a pointer to some of the means by which benefits in health and safety can be delivered.

I have been fascinated over the last six years as I witnessed the impact on the construction industry of what was a radical approach to health and safety management. This second edition has been able to affirm many of the predictions that were to be found in the first edition. My thanks, therefore, are due to all those persons who have been involved in the application of the Regulations and those who have contributed to, and enriched, the 'CDM debate'.

Raymond Joyce

June 2001

Contents

Table of cases

Table of UK statutes

Civil Evidence Act 1995
Company Directors Disqualification Act 1986
Dentists Act 1984
European Communities Act 1972
Factories Act 1961
Health and Safety at Work etc. Act 1974
Interpretation Act 1978
Occupier's Liability Act 1957
Occupier's Liability Act 1984
Single European Communities (Amendment) Act 1986

Table of statutory instruments

1991 Public Works Contracts Regulations (SI 1991: 2680)

1992 Provision and Use of Work Equipment Regulations (SI 1992: 2932)

1992 Personal Protective Equipment at Work Regulations (SI 1992:2966)

1992 Personal Protective Equipment (EC Directive) Regulations (SI 1992: 3139)

1993 Personal Protective Equipment (EC Directive) (Amendment) Regulations (SI 1993: 3074)

1994 Personal Protective Equipment (EC Directive) (Amendment) Regulations (SI 1994: 2326)

1996 Personal Protective Equipment (EC Directive) (Amendment) Regulations (SI 1996: 3039)

1956 Quarries Order (SI 1956: 1780)

1995 Reporting of Injuries, Diseases and Dangerous Occurrences Regulations (SI 1995: 3163)

1977 Safety Representatives and Safety Committees Regulations (SI 1977: 500)

1992 Workplace (Health, Safety and Welfare) Regulations (SI 1992: 3004)

1994 Waste Management Licensing Regulations (SI 1995: 3163)

Table of EU legislation

Treaty of Rome 1957
Article 118A
Directive 89/391/EEC
Directive 89/654/EEC
Directive 89/655/EEC
Directive 89/656/EEC
Directive 90/269/EEC
Directive 90/270/EEC
Directive 92/85/EEC
Directive 93/95/EEC
Directive 93/68/EEC
Directive 92/57/EEC

Table of abbreviations

AC	Appeal cases
the **ACOP**	the Approved Code of Practice
All ER	All England Law Reports
Amending Regulations	The Construction (Design and Management) (Amendment) Regulations 2000
BLM	Building Law Monthly
the **ChD**	Chancery Division
the **C(HSW) Regulations**	The Construction (Health, Safety and Welfare) Regulations 1996
CMLR	Common Market Law Reports
the **Commission**	the Health and Safety Commission
ECR	European Court Reports
the **EU**	the European Union
the **Executive**	the Health and Safety Executive
the **Framework Directive**	Council Directive of 12th June 1989 on the introduction of measures to encourage improvements in the safety and health of workers at work (89/391/EEC)
the **Framework Regulations**	the Management of Health and Safety at Work Regulations 1999
HMSO	Her Majesty's Stationery Office
HSWA 1974	the Health and Safety at Work etc. Act 1974
ICE Conditions	ICE Conditions of Contract, Seventh Edition
JCT 98	JCT Standard Form of Building Contract, 1998 Edition

LR Ex	Law Reports Exchequer
the **Manual Handling Regulations**	the Manual Handling Operations Regulations 1992
the **PPE Regulations**	the Personal Protective Equipment Regulations 1992
the **Regulations**	the Construction (Design and Management) Regulations 1994
the **Single European Act**	the Single European Communities (Amendment) Act 1986
the **Temporary or Mobile Worksites Directive**	Council Directive of 25th June 1992 on the implementation of minimum safety and health requirements at temporary or mobile construction sites (92/57/EEC)
the **VDU Regulations**	the Health and Safety (Display Screen Equipment) Regulations 1992
the **Workplace Regulations**	the Workplace (Health, Safety and Welfare) Regulations 1992
the **Work Equipment Regulations**	the Provision and use of Work Equipment Regulations 1992

1 Introduction

The facts
The analysis
Towards an improvement
The solution?
Costs and benefits

The facts

The construction industry's safety record continues to confirm the industry as one of the most dangerous. In 1988 the Executive published a study of accidents over five years in the building and civil engineering industries entitled, 'Blackspot Construction'. It reported that in the five year period of 1981–1985, 739 people were killed in the construction industry.

The grim statistics since 1985 continue to demonstrate that the construction industry has become no less dangerous. Figure 1 illustrates the number of fatalities each year throughout the period from 1981 to 2000. Until the depth of the recession and a fall in workload in the industry during 1991/92, fatalities had been remarkably constant, at an average of 136 each year. As the economy recovered from the recession, it can be seen that the fatalities have reached a reasonably stable level but without any hint of future improvements. The conclusion behind such a connection between construction activity levels and fatalities is that health and safety is still not a sufficient priority to attract investment in training and the complacent mantra that 'accidents will happen' still rules.

The number of fatalities in the construction industry are only the tip of the iceberg. There are thousands of major injuries each year and even more minor injuries which result in an absence from normal work on more than three consecutive working days. The Executive knows that only a fraction of the non-fatal injuries are reported each year. The 1990 Labour Force Survey estimated that only 40% of reportable non-fatal injuries to employees within the construction sector are reported.

1

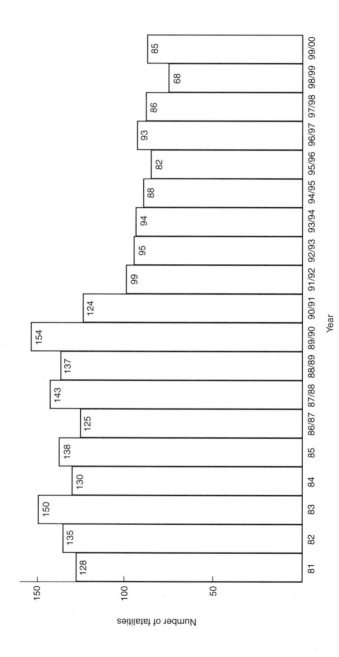

Figure 1   Fatalities in the construction industry since 1981 (from 1986 the reporting year is April to March)

The analysis

The Executive concluded in 'Blackspot Construction' that 70% of the deaths could have been prevented by positive action by managers within the industry. It was the Executive's opinion that a majority of the accidents could be prevented by the application of reasonably practicable precautions. The analysis of the main causes of accidents by the Executive revealed that most were due to the following reasons.

1. A lack of supervision by line managers in the industry. This was identified by the Executive, and is a view supported by the unions. It was felt that the widespread use of sub-contractors and self-employed labour led to problems of management and control, which were exacerbated by new forms of contracting which involved management remote from the site.

2. Custom and practice in the industry was not equipping workers to identify dangers and take steps to protect themselves.

3. A lack of co-ordination between the members of the professional team at the pre-construction stage. This was revealed by a study of the overall management of construction sites.

    The preamble to the directive on the implementation of minimum safety and health requirements at temporary or mobile sites ('the Temporary or Mobile Worksites Directive') concurred in recognising the importance of the pre-construction phase as a cause of accidents across the European Union ('the EU'), as follows:

    Whereas unsatisfactory and/or organisational options or poor planning of the works at the project preparation stage have played a role in more than half of the occupational accidents occurring on construction sites in the community.

Towards an improvement

The concern of the Health and Safety Commission ('the Commission') and the Executive continued unabated in the late 1980s, which resulted in the Commission publishing a consultative document in September 1989 entitled 'Construction Management – Proposals for Regulations An Approved Code of Practice'. The consultative document proposed regulations and an ACOP for construction management and declared its intention to improve health and safety in the construction industry.

Earlier in 1989 the European Community (as it then was) agreed and adopted a number of directives for implementation by 31 December 1992. These directives and the regulations which implement them, known colloquially as the 'six-pack' are discussed in Chapter 2. All the regulations are applicable to the construction industry, with the exception of the Workplace (Health, Safety and Welfare) Regulations, which expressly exclude construction sites and activities.

The EU declared its intention to produce a separate directive for the construction industry. This was to be known as the Temporary or Mobile Worksites Directive. In view of this declared intention, the British government did not proceed with introducing new legislation based on the consultative document published in September 1989. However, the comments which were received in response to the consultative document were used as the basis for the British government's negotiations on the final format of the directive.

Finally, the Temporary or Mobile Worksites Directive, which was agreed and adopted on 24 June 1992, required Member States to implement its terms into national legislation by 1 January 1994.

The solution?

The Commission published in September 1992 a consultative document entitled 'Proposals for Construction (Design and Management) Regulations and Approved Code of Practice' which set out the proposals for implementing the Temporary or Mobile Worksites Directive in Great Britain. The document initiated an extended period of public consultation, which resulted in various changes to the original proposal before culminating in the laying before Parliament of the Regulations on 10 January 1995.

The Regulations are not prescriptive; they avoid setting standards. Emphasis is placed on identifying hazards and the assessment of risk. The members of a project team are required to act upon the risk assessment so as to eliminate, avoid or (at the least) lessen perceived risks. The influence of the Framework Regulations is particularly important in this regard. The duties set out in the Regulations were intended to improve the management of construction projects from inception through to completion, and therefore all the members of a project team are subject to the Regulations.

The Regulations have not, in the opinion of many in the construction industry, delivered the benefits in improving health and safety standards that had been expected. The opposing view points to the fact that parties to a construction project have considered health and safety at the outset and, therefore, how can there not be an improvement. Even the Executive has expressed doubts about the impact of the Regulations and, yet, who would dare to evaluate how many more fatalities and injuries there would have been without the Regulations?

There has undoubtedly been a great amount of unproductive and inefficient administration that has detracted from the benefit of the Regulations. The excesses of the administrative procedures have been redressed to some extent by experience, and the longer-term benefit of the Regulations will create a greater awareness of safety culture. Nonetheless, the Executive have found inconsistencies in the way that some duty holders have interpreted their responsibilities. The bigger issue for legislation, which is not prescriptive, is the need to have individuals taking responsibility for health and safety who have the requisite intelligence, experience, training and lateral thought.

Cost and benefits

The Executive have estimated that the total economic cost to Great Britain of employers and other duty holders failing to comply with health and safety requirements is up to £18 billion each year. On the basis that the construction industry contributes more than any other to the appalling statistics, there is huge scope for making savings by investing in health and safety.

The Executive acknowledged when the Regulations were first introduced that there would be a significant cost to organisations that did not already have a well-developed approach to health and safety management.

In the consultative document for the Regulations, the Commission assumed (whilst acknowledging the difficulties) that the total cost to the construction industry in implementing the Regulations would be in the region of £550 million, based on an industry output of £37,000 million for 1991. The two main areas of additional cost were associated with the duties upon designers and the duty upon the planning supervisor and principal contractor to produce a health and safety plan. The Commission estimated that compliance by the designers with their new duties might cost up to an additional £290 million each year, and by the planning supervisor and principal contractor an additional £185 million each year.

The Commission recognised that the reduction in the level of accidents would be the principal quantifiable benefit. They assumed that on small to medium-sized sites, the reduction in accidents would be 33% if the Regulations were implemented, whereas on large sites, where safety management is usually better developed, a 20% reduction in accidents could be expected. The Commission concluded that the estimated benefit to the industry would be £220 million each year.

No one can know what the outturn figures have been for costs and improvements in the accident rate – the cost benefit equation will remain imprecise and more instinctive than certain. However, there is evidence that good health and safety practice brings improved productivity and product quality, which can contribute to increased profitability for industry.

There is no doubt that accidents on construction sites lead to disruption, delays and sometimes redesign but it is precisely because the cost benefit equation is imprecise that the construction industry has not yet, perhaps, fully learned how to apply the Regulations to deliver the intended benefits.

2   Framework of health and safety law

Introduction
Common law
  'Reasonably practicable'
Statute law
  The Factories Act 1961
  Health and Safety at Work etc. Act 1974
  European legislation

Introduction

The Regulations are but a small part of a very much larger legal framework of law which affects matters of health and safety.

The Regulations, except for some minor repeals, revocations and modifications set out in regulation 24, do not change the law. They are, however, a significant and important part of the health and safety and general regulatory approach to the construction industry.

Common law

Common law is often referred to as that part of English law which cannot be found in Acts of Parliament but in the decision of the courts and custom. Common law principles change and develop with the decisions of the courts. Conversely, Acts of Parliament can abolish well-established rules of common law.

The importance of common law in health and safety matters has diminished considerably with the increased amount of legislation. Statute law will always prevail whenever there is a conflict with the common law, thus upholding the legislative supremacy of Parliament. However, despite the extensive legislation which affects matters of health and safety, common law principles co-exist with statutory provision. For example, there is implied into every contract of employment a term that the employer will provide safe plant and premises, a safe system of work and reasonably competent fellow workers.

The common law, insofar as it is based on decisions of the courts, also serves to give effect to the intentions of Parliament in the interpretation of Statutes. In legislation concerned with health and safety matters, including the Regulations, the expression 'reasonably practicable' is used widely.

'Reasonably practicable'

The case of *Edwards -v- National Coal Board* was concerned with the duty of a mine owner to support the mine roof. The guiding principle as to the interpretation of 'reasonably practicable' in that case was stated by Asquith LF as follows:

> 'Reasonably practicable' is a narrower term than 'physically possible' and seems to me to imply that a computation must be made by the owner in which the quantum of risk is placed on one scale and the sacrifice involved in the measures necessary for averting the risk (whether in money, time or trouble) is placed in the other, and that, if it is shown that there is a gross disproportion between them – the risk being insignificant in relation to the sacrifice – the defence has discharged the onus on them. Moreover, this computation falls to be made by the owner at a point in time and anterior to the accident. The questions he has to answer are:
>
> (a) What measures are necessary and sufficient to prevent any breach ...?; and
>
> (b) are these measures reasonably practicable?

Therefore, to determine what is reasonably practicable a cost benefit exercise has to be carried out. If the risk is insignificant in relation to the sacrifice it is not reasonably practicable to take steps necessary to control a hazard.

Based upon the above judgment 'Croners Health & Safety at Work' gives the following definition of reasonably practicable:

> The duty to do what is reasonably practicable is less strict than the unqualified duty to do what is 'practicable'. The introduction of the qualifying word 'reasonably' implies that a computation must be made in which the question of risk is placed on one side and the sacrifice involved in instituting the measures necessary for alleviating the risk (whether in money, time or trouble) is placed on the other. If, when this

'cost benefit' exercise has been carried out, the risk is insignificant in relation to the sacrifice, it is not reasonably practicable to take the steps necessary to control the hazard.

Thus, an employer or duty holder who conducts a cost benefit analysis and concludes that the costs of implementing health and safety procedures outweigh the benefit is, in theory, not obliged to take any further action. But a word of warning, as a basis for a defence such arguments are not convincing except in the most compelling circumstances.

Statute law

Numerous Acts of Parliament and statutory instruments affect the construction industry. Although for the purposes of this book there are too many to be mentioned, those affected by the Regulations, or of outstanding significance, are referred to below to set the statutory framework within which they exist.

The Factories Act 1961

The Factories Act 1961 used to be the starting point for consideration of Parliamentary legislation on health and safety matters in the construction industry. However, regulation 24(1) of the Regulations repeals subsections (6) and (7) of section 127 of the Factories Act which deal with notification to the Executive and the impact of the 'six-pack' (see below) and the Construction (Health, Safety and Welfare) Regulations 1996 has only left remnants of the Factories Act with any relevance.

Regulation 24(2) of the Regulations amended the Construction (General Provisions) Regulations 1961 made under the Factories Act, although, these regulations, together with the Construction (Health and Welfare) Regulations 1996 and the Construction (Working Places) Regulations 1966, have now been replaced by the Construction (Health, Safety and Welfare) Regulations 1996. Also, the Construction (Lifting Operations) Regulations 1961 have been replaced by the Lifting Operations and Lifting Equipment Regulations 1998.

The Construction (Notice of Operations and Works) Order 1965 made under The Factories Act is revoked by regulation 24(3) of the Regulations.

Health and Safety at Work etc. Act 1974

The Health and Safety at Work etc. Act 1974 (HSWA 1974) established the Health and Safety Commission ('the Commission') and gave it power to

propose health and safety regulations and approve codes of practice. It also
set up the Health and Safety Executive ('the Executive') with responsibility for enforcing health and safety laws.

The radical difference between HSWA 1974 and all preceding health
and safety legislation was the emphasis on individuals and their duties as
compared with premises. Rather than a prescriptive approach, HSWA
1974 is based on principles designed to bring about a greater awareness of
the problems associated with health and safety issues. The legislation is
addressed to individuals and includes employers, employees and producers
of industrial products.

As the primary safety legislation in the UK, HSWA 1974 is the Act
under which virtually all subsequent health and safety regulations have
been made. For example, the 'six-pack' regulations which are referred to
below, and the Regulations are all made under HSWA 1974.

The Regulations are made under HSWA 1974. The Regulations do not
change HSWA 1974 or repeal any part of it although regulation 24(4)
makes a minor modification to the Health and Safety (Enforcing
Authority) Regulations 1989 which were made under HSWA 1974.
Regulation 24(4) provides:

> *For item (i) of paragraph 4(a) of Schedule 2 to the Health and
> Safety (Enforcing Authority) Regulations 1989, the following item
> shall be substituted –*

> *"(i)   regulation 7(1) of the Construction (Design and Management)
>   Regulations 1994 (S. 1994/3140) (which requires projects
>   which include or are intended to include construction work to
>   be notified to the Executive) applies to the project which
>   includes the work; or"*

European legislation

The European Communities Act 1972, which came into effect from 1
January 1973, incorporated the Treaties constituting the European
Economic Community, now called the European Union (EU), into United
Kingdom legislation. EU law, which includes Directives, becomes law in
the United Kingdom by various means, notably with regard to health and
safety legislation, by Statutory Instrument.

The European health and safety initiative
The Single European Act introduced Article 118A into the Treaty of Rome
which obliges the Member States to 'pay particular attention to encouraging

improvements, especially in the working environment, as regards the health and safety of workers'. Article 118A also provides that the health and safety policy would be introduced, by the adoption of Directives on working conditions and technical standards, on the basis of majority voting in the EU Council of Ministers. The Directives adopted under Article 118A are intended to avoid, as far as possible, imposing administrative, financial and legal constraints that would 'hold back the creation and development of small and medium-sized undertakings'. The Directives are only intended to establish minimum standards. Member States are not prevented from maintaining or introducing more stringent measures, providing the measures do not interfere with other objectives of the Treaty of Rome, including measures which would have an anti-competitive influence on the open market. To a greater or lesser extent, all the Directives adopted under Article 118A and the regulations which implement them, have an impact on the construction industry, and share the same overall strategic objective with the Regulations which form the subject matter of this book: to improve the health and safety of workers.

The Directives which have been adopted under Article 118A include:

(i) Council Directive of 12 June 1989 on the introduction of measures to encourage improvements in the health and safety of workers at work (89/391/EEC);

(ii) Council Directive of 30 November 1989 concerning the minimum safety and health requirements for the workplace (first individual directive within the meaning of Article 16(1) of Directive 89/391/EEC)(89/654/EEC);

(iii) Council Directive of 30 November 1989 concerning the minimum safety and health requirements for the use of work equipment by workers at work (second individual directive within the meaning of Article 16(1) of Directive 89/391/EEC)(89/655/EEC);

(iv) Council Directive of 30 November 1989 on the minimum health and safety requirements for the use by workers of personal protective equipment at the work place (third individual directive within the meaning of Article 16(1) of Directive 89/391/EEC)(89/656/EEC);

(v) Council Directive of 29 May 1990 on the minimum health and safety requirements for the manual handling of loads where there is a risk particularly of back injury to workers (fourth individual directive within the meaning of Article 16(1) of Directive 89/391/EEC)(90/269/EEC);

(vi) Council Directive of 29 May 1990 on the minimum safety and health requirements for work with display screen equipment (fifth individual directive within the meaning of Article 16(1) of Directive 89/391/EEC)(90/270/EEC);

(vii) Council Directive of 24 June 1992 on the implementation of minimum safety and health requirements at temporary or mobile construction sites (eighth individual directive within the meaning of Article 16(1) of Directive 89/391/EEC)(92/57/EEC).

Implementation into domestic law

Preceding the Regulations there were six other regulations which have become known as the 'six-pack', made under section 15 of HSWA 1974 (see Figure 2). The six-pack enacted the first tranche of a family of Directives adopted under Article 118A. The six-pack regulations are summarised briefly as follows:

1.   *The Management of Health and Safety At Work Regulations 1999 (The Framework Regulations)*

     The Management of Health and Safety at Work Regulations 1992, which came into effect on 1 January 1993, implemented EC Directive 89/391/EEC otherwise known as the Framework Directive. The regulations have since been re-enacted by the Management of Health and Safety at Work Regulations 1999.

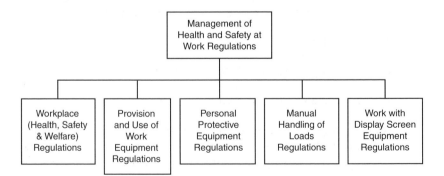

Figure 2   The 'six-pack' regulations

The Framework Directive is so called because it creates a framework for further Directives. By creating broad and general duties on employers, employees and the self-employed, the need for dealing with specific hazards and work situations was dealt with in subsequent Directives.

Certain obligations contained in the Framework Directive were already to be found in domestic legislation, including HSWA 1974 and the Safety Representatives and Safety Committees Regulations 1977. Accordingly, the Framework Regulations were drafted to avoid repeals or revocation of any existing domestic legislation.

The Framework Regulations, which cover all areas of work, except sea transport, apply to all businesses regardless of size. However, the nature, extent and cost of the measures which need to be taken to ensure compliance will vary, depending on the nature of the hazards and level of risks associated with different industries.

The three main requirements of the Framework Regulations concern:

1. risk assessment;

2. arrangements for protective and preventative measures; and

3. the appointment of competent persons to assist with protective and preventative measures.

The re-enacted Framework Regulations have had the effect of numerous consequential amendments to other regulations but more significantly an employer is now required to implement any preventative and protective measures on the basis of the following principles set out at Schedule 1 of the Framework Regulations:

(a) avoiding risks;

(b) evaluating the risks that cannot be avoided;

(c) combating the risks at source;

(d) adapting the work to the individual, especially as regards the design of workplaces, the choice of work equipment and the choice of working and production methods, with a view, in particular, to alleviating monotonous work and work at a predetermined work rate and to reducing their effect on health;

(e) adapting to technical progress;

(f) replacing the dangerous by the non-dangerous or the less dangerous;

(g) developing a coherent overall prevention policy that covers technology, organisation of work, working conditions, social relationships and the influence of factors relating to the working environment;

(h) giving collective protective measures priority over individual protective measures; and

(i) giving appropriate instructions to employees.

2. *The Workplace (Health, Safety and Welfare) Regulations 1992 (The Workplace Regulations)*
    The Workplace (Health, Safety and Welfare) Regulations 1992 which came into effect on 1 January 1993 implement EC Directive 89/654/EEC. Temporary and mobile sites were excluded specifically from this Directive.
    The Workplace Regulations do not apply to ships, building operations, works of engineering construction, mines, quarries or other mineral extraction. The ACOP provides guidance on the application of the Workplace Regulations and affirms that construction sites (including site offices) are excluded. Construction work within a workplace will be treated as a construction site if it is fenced off and therefore excluded from the Workplace Regulations. Conversely, the requirements of the Workplace Regulations for sanitary conveniences, washing facilities, drinking water, clothing, accommodation, changing facilities and facilities for rest and eating meals apply 'so far as is reasonably practicable' to temporary sites.

3. *The Provision and Use of Work Equipment Regulations 1992 (The Work Equipment Regulations)*
    The Provision and Use of Work Equipment Regulations 1992 which came into force on 1 January 1993 implement EC Directive 89/655/EEC. There was a transition period for work equipment first provided for use before 1 January 1993 such that the entire Work Equipment Regulations did not have full effect until 1 January 1997.
    The Work Equipment Regulations impose duties on employers, persons in control of certain premises and the self-employed to

ensure the provision of suitable equipment, maintenance, proper training, information and instruction.

Work equipment is defined in general terms to include any machinery, appliance, apparatus or tool. Work equipment on a construction site could include everything from a pick and wheelbarrow to a concrete paving train.

There is no ACOP for the Work Equipment Regulations although the Executive have issued Guidance Notes.

4. *The Personal Protective Equipment at Work Regulations 1992 (The PPE Regulations)*
   The PPE Regulations which came into full force on 1 January 1993 implement EC Directive 89/656/EEC.

   The PPE Regulations do not affect any of the pre-existing regulations or other statutory requirements which are concerned with the personal protection of employees such as the noise and head protection regulations. However the PPE Regulations impose a duty on all employers to assess and deal with those specific risks which were not already regulated. Thus, in the construction industry construction contractors are required, amongst other matters, to assess the hazards to health from working in inclement weather conditions and provide the appropriate weather protection.

   There is no ACOP for the PPE Regulations, although the Executive have issued Guidance Notes.

   All personal protective equipment used in the European Union must satisfy or comply with applicable standards that are set out in the Personal Protective Equipment (EC Directive) Regulations 1992, as amended by the Personal Protective Equipment (EC Directive) (Amendment) Regulations 1993, the Personal Protective Equipment (EC Directive) (Amendment) Regulations 1994 and the Personal Protective Equipment (EC Directive) (Amendment) Regulations 1996. Together, all these regulations seek to ensure that all personal protective equipment for the workplace complies with essential safety requirements. The 'CE' mark is evidence that such equipment complies with the regulations.

5. *The Manual Handling Operations Regulations 1992 (The Manual Handling Regulations)*
   The Manual Handling Regulations came into force, and with full effect, on 1 January 1993 and implement Directive 90/269/EEC.

The Manual Handling Regulations apply to all types of occupation except to the narrow category of a master or crew on sea-going ships. Manual handling operations are defined as 'any transporting or supporting of a load (including the lifting, putting down, pushing, pulling, carrying or moving thereof) by hand or by bodily force'.

The construction industry relies to a significant extent on manual handling and has had to pay particular attention to the Manual Handling Regulations.

The Manual Handling Regulations do not cover injuries caused by any leakage or spillage of any toxic or corrosive substance from a load but are concerned with injuries such as muscular strains and fractures. Employers have to identify hazards and risks associated with normal work activities and take the appropriate action to remove or reduce the risk of injury.

There is no ACOP for the Manual Handling Regulations, although the Guidance Notes produced by the Executive are particularly detailed.

6. *The Health and Safety (Display Screen Equipment) Regulations 1992 (The VDU Regulations)*
    The VDU Regulations which came into force on 1 January 1993 implement EC Directive 90/270/EEC. The full effect of these regulations will apply from 1 January 1997 when the exemption for workstations in use prior to 1 January 1993 has expired.

    The use of VDU workstations within offices on construction sites is not commonplace. Despite the view of many in the construction industry that these regulations are of least influence when compared with the others in the 'six-pack', the design and management of construction projects relies heavily on computer applications and therefore they should not be overlooked.

    There is no ACOP although the Executive have produced Guidance Notes.

3 An overview of the Regulations

Introduction

The Regulations give effect to Council Directive 92/57/EEC on the
implementation of minimum safety and health requirements at temporary
or mobile construction sites. Regulation 1 provides:

> *These Regulations may be cited as the Construction (Design and
> Management) Regulations 1994 and shall come into force on 31st
> March 1995.*

To what extent do the Regulations give effect to the Temporary or Mobile Worksites Directive?

The Temporary or Mobile Worksites Directive, which is reproduced as Appendix 1, is not implemented in its entirety by the Regulations. Parts of Article 10, which deals with self-employed persons on construction sites, were already implemented by the 'six-pack'.

1.   The Management of Health and Safety at Work Regulations implement the first part of Article 10.1(a)(i) and Article 10.2(a)(i).

2.   The Provision and Use of Work Equipment Regulations implement, where earlier regulations had not already done so, Articles 10.1(a)(ii) and 10.2(a)(ii).

3.   The Use of Personal Protective Equipment Regulations implement Articles 10.1(a)(iii) and 10.2(a)(iii).

The Regulations implement a substantial part of the remainder of the Temporary or Mobile Worksites Directive except for the following (in addition to items 1–3 above):

(a)   Sub-paragraphs (a)–(h) and (j) of Article 8 in so far as these particulars are not required by regulation 15 to be included in the health and safety plan.

(b)   Sub-paragraph (a) of Article 9.

(c)   Generally, in so far as the Regulations do not apply to:

   (i)   projects (other than for demolition or dismantling of a structure) in respect of which no more than four persons are carrying out construction work at any one time; and

   (ii)   minor construction work in respect of which the Executive is not the enforcing authority; and

   (iii)   construction work forming part of a project carried out for a domestic client except where regulation 5 applies.

Construction (Health, Safety and Welfare) Regulations 1996

Annex IV of the Directive has been implemented by the Construction (Health, Safety and Welfare) Regulations 1996, which came into force on

2 September 1996. These regulations made a small amendment to the definition of 'construction work' to incorporate the Carriage of Dangerous Goods by Road and Rail (Classification, Packaging and Labelling) Regulations 1994.

The Management (Health, Safety and Welfare) Regulations also provide an informative checklist for the risk assessment necessary for the Health and Safety Plan, which is discussed in Chapter 10.

The Amending Regulations

The Amending Regulations, which concern regulations 2, 12 and 13 were hurriedly introduced after the Court of Appeal decision in *R. -v- Wurth SA* revealed a loophole in the application of the Regulations to designers. The Amending Regulations were generally uncontroversial, although they are the only substantive amendments when compared with the consequential amendments brought about by changes in other legislation. A full review of the Amending Regulations is set out in Chapter 6, which deals with the designer.

Outline of the principal features of the Regulations

The Regulations are concerned with projects which are defined in regulation 2(1) which provides:

> *'project' means a project which includes or is intended to include construction work.*

The Regulations bring health and safety management, on an obligatory basis, into the planning and design of construction work, of all but the smallest projects, which are exempted from the Regulations. Exempted projects are, of course, still subject to the full requirements of health and safety legislation generally. No longer is the contractor left with the sole responsibility for health and safety during construction, for under the Regulations, it is shared with the other parties to a project. The Regulations direct how the parties to a project will contribute to health and safety management and therefore sets out the key roles as follows.

The client

The client has an obligation to appoint a planning supervisor and principal contractor. The client must be satisfied that the persons so appointed to those roles are competent and that they have allocated or will allocate sufficient resources, including time, to the project. Whilst the client is not obliged to appoint designers, in circumstances where the client does make such an appointment the requirement to be satisfied with regard to competence and the allocation of sufficient resources also applies. Only domestic householders, acting as clients, procuring construction work carried out on their domestic premises are exempted from the Regulations.

The client can appoint an agent who will be treated for the purposes of the Regulations as the client, subject to the Executive receiving the appropriate notice.

The planning supervisor

One of the most significant changes brought about by the Regulations to the composition of a project team was the introduction of the planning supervisor, which was a wholly new role. The appointment of a planning supervisor is required on all projects, other than those exempted from the Regulations. The planning supervisor has to ensure that a health and safety plan is prepared, monitor the health and safety aspects of the design, be in a position to give adequate advice to the client and any contractor and ensure a health and safety file is prepared in respect of the project.

The planning supervisor has become an important source of advice to clients with regard to the Regulations.

The designer

The designer is required to make a client aware of the duties under the Regulations to be performed by the client. The designer is also required to ensure that the design avoids unnecessary risks to health and safety or reduces the risks so that the projects, for which they have designed, can be constructed and maintained safely.

The principal contractor

The principal contractor is required to take over and develop the health and safety plan, co-ordinate the activities of other contractors, in addition to the duties in relation to providing information, training and consultation with employees, including the self-employed.

Other contractors

Contractors are required to co-operate with the principal contractor. This includes complying with directions given by the principal contractor and providing him with details on the management and prevention of health and safety risks created by the contractors' work on site. Contractors contribute to the management of health and safety on site by the provision of other information to the principal contractor and employees.

Each of the above roles is discussed in detail under the chapters identified by the relevant title.

Competence and adequate resources

The Regulations stress the need for the persons fulfilling the different roles, except for the role of client, to be competent. The Regulations also introduce a requirement for the same persons to demonstrate that they have allocated or intend to allocate adequate resources to health and safety. A detailed discussion of competence and the allocation of adequate resources can be found in Chapter 4.

Health and safety plan

A co-ordinated approach to the health and safety management of the construction project relies upon communication between the participants. The health and safety plan, which is an innovative feature of the Regulations, is the mechanism which binds the participants together, to improve the exchange and communication of matters affecting health and safety. During the pre-construction phase, the health and safety plan is prepared on information obtained from the client and designers, and where appropriate, the planning supervisor. Before the construction phase can begin the health and safety plan must be developed by the principal contractor. He does this by including details on the management and prevention of health and safety risks created by contractors and sub-contractors. The health and safety plan is a dynamic document subject to continuous review and amendment, fulfilling its role as a co-ordinating mechanism, as construction progresses.

Health and safety file

On the completion of a construction project the planning supervisor is required to ensure that a health and safety file is prepared and handed over to the client. The health and safety file is an important record document hould be easily available to others responsible for later construction associated with the structure or its maintenance, repair, renovation or demolition.

The Approved Code of Practice ('The ACOP')

The ACOP, approved by the Health and Safety Commission, with the consent of the Secretary of State for Employment under section 16(1) of the HSWA 1974 provides practical guidance on compliance with the Regulations. Failure to comply with the ACOP is not in itself an offence, although such failure may be taken by a court in criminal proceedings as proof that a person has contravened the Regulations. In such circumstances however, it is open to a person to satisfy the court that he has complied with the Regulations in some other way.

The dependence on the ACOPs or guidance notes for the Regulations and the 'six-pack' are, in the view of some writers, quantitatively different from the traditional role of 'fleshing out' certain sets of regulations. Whilst the Regulations are fundamental, it may be that in practice it is the extensive guidance in the ACOP that assumes the dominant role as to what is or is not required. No person engaged in any of the roles identified in the Regulations can afford to ignore the guidance provided by the ACOP.

The Executive has sought and obtained comments on proposed changes to the present ACOP. A revised ACOP is likely to be published in 2002, although there is no reliable indication when that might be.

Scope of application

Regulation 3(1) provides that:

> *Subject to the following paragraphs of this regulation, these Regulations shall apply to and in relation to construction work.*

It is vital to appreciate the definition of construction work, because any activity which is not within the definition of 'construction work' is not subject to the Regulations.

The definition of 'construction work' in Regulation 2(1) as amended by the Construction (Health, Safety and Welfare) Regulations 1996 is widely drawn as follows:

the carrying out of any building, civil engineering or engineering construction work and includes any of the following –

(a)   *the construction, alteration, conversion, fitting out, commissioning, renovation, repair, upkeep, redecoration or other maintenance (including cleaning which involves the use of water or an abrasive at high pressure or the use of substance classified as corrosive or toxic for the purposes of Regulation 5 of the Carriage of Dangerous Goods by Road and Rail (Classification, Packaging and Labelling) Regulations 1994 [SI 1994/669] de-commissioning, demolition or dismantling of a structure;*

(b)   *the preparation for an intended structure including site clearance, exploration, investigation (but not site survey) and excavation, and laying or installing the foundations of the structure;*

Note that 'site survey' is excluded and so is not required in the health and safety plan, despite the fact it is an activity which could be construed in accordance with regulation 15(5)(b).

(c)   *the assembly of prefabricated elements to form a structure or the disassembly of prefabricated elements which, immediately before such disassembly, formed a structure;*

The ACOP suggests that if the work described in regulation 2(1)(c) is carried out in factories or off-site workshops making products which are used subsequently in connection with construction work, the Regulations will not apply, except with regard to designers. However, it is suggested that it will, in some cases, be a matter of degree whether or not the Regulations apply. For example, the trial erection of a steel girder bridge in a steel fabrication workshop may be of such size and complexity as to represent a construction project in its own right. Whilst the Workplace Regulations might apply, the need to identify the hazards and risks associated with such an exercise could only properly be dealt with in accordance with the Regulations.

(d)   *the removal of a structure or part of a structure or of any product or waste resulting from demolition or dismantling of a structure or from disassembly of prefabricated elements which, immediately before such disassembly, formed a structure;*

Construction work falling within this sub-definition is always subject to the Regulations due to the high risk of personal injury.

> *(e)* *the installation, commissioning, maintenance, repair or removal of mechanical, electrical, gas, compressed air, hydraulic, telecommunications, computer or similar services which are normally fixed within or to a structure;*

The degree of fixity is not described nor elaborated upon in the ACOP. It is suggested that actual connection is required, as opposed to a free-standing item, e.g. a computer 'fixed' by its own weight

> *but does not include the exploration for or extraction of mineral resources or activities preparatory thereto carried out a place where such exploration or extraction is carried out.*

Construction work, as defined, associated with the exploration for or extraction of mineral resources is excluded specifically, as is the setting up of any infrastructure for such exploration or extraction. Thus, all work concerned directly with mineral exploration or extraction including deep and open cast coal mining, clay pits, sand, stone and aggregate extraction, is expressly excluded. The ACOP highlights the difference between oil rigs and similar equipment, referred to as articles, and other structures defined within the Regulations. Only construction work and structures directly concerned with exploration or extraction are exempted. Thus, the Regulations do apply to building, civil engineering or engineering construction work at quarries and mines associated with the development of a site prior to operations to extract mineral resources. They also apply to operations or mines and quarries which are not directly related to the work to extract mineral resources

The definition of construction work cannot be appreciated fully without a knowledge of the term 'structure', which is defined as:

> *(a)* *any building, railway line or siding, tramway line, dock, harbour, inland navigation, tunnel, shaft, bridge, viaduct, waterworks, reservoir, pipe or pipe-line (whatever, in either case, it contains or is intended to contain), cable, aqueduct, sewer, sewage works, gas holder, road, air field, sea defence works, river works, drainage works, earthworks, lagoon, dam, wall, caisson, mast, tower, pylon, underground tank, earth retaining structure, all structures designed to preserve or alter any natural feature, and any other structure similar to the foregoing;*

(b) *any formwork, falseworks, scaffold or other structure designed or used to provide support or means of access during construction work;*

(c) *any fixed plant in respect of work which is installation, commissioning, de-commissioning or dismantling and where any such work involves a risk of a person falling more than two metres.*

Regulation 3(2) provides:

Subject to paragraph (3), Regulations 4 to 12 and 14 to 19 shall not apply to or in relation to construction work included in a project where the client has reasonable grounds for believing that –

(a) *the project is not notifiable; and*

(b) *the largest number of persons at work at any one time carrying out construction work included in the project will be or, as the case may be, is less than 5*

To ascertain whether a project is notifiable, regulation 2(4) states the following:

For the purposes of these Regulations, a project is notifiable if the construction phase –

(a) *will be longer than 30 days; or*

(b) *will involve more than 500 person days of construction work;*

and the expression 'notifiable' shall be construed accordingly.

To determine whether regulations 4 to 12 and 14 to 19 apply, a client has to decide whether or not the project is notifiable before being committed to appointing a planning supervisor or principal contractor in accordance with regulation 6. The ACOP confirms that a working day is any day on which any construction work is carried out, even if the work takes less than a day or the day is a holiday or over a weekend. A person day is one individual, including supervisors and specialists, carrying out construction work for one normal working shift. Therefore, it should be assumed that a visiting clerk of works, engineer or architect for example, is counted as a person for the purposes of regulation 3(2)(b). Figure 3 illustrates how the

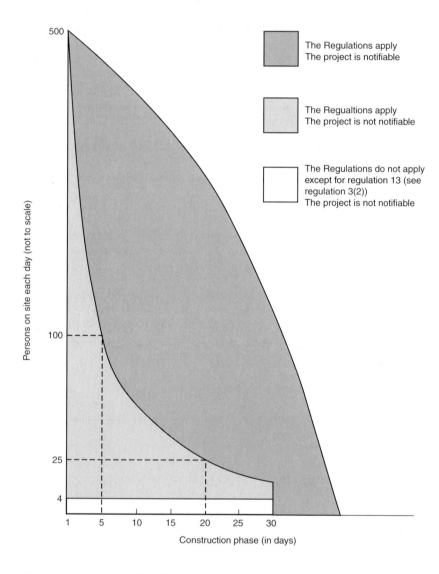

Figure 3 A diagram (for illustrative purposes only) showing how the number of persons on site and the duration of the construction phase affect notification and the application of the Regulations

number of persons on site and the duration of the construction phase affects notification and the application of the regulations.

Determining the resources and timescales required for completing construction work can involve difficult matters of technical and professional judgment. Without adequate in-house resources or the services of a construction professional how is a client able to evaluate that 'reasonable grounds' had been established to take advantage of the exclusion of regulations 4 to 12 and 14 to 19? In a situation where there is any doubt as to the reliability of establishing reasonable grounds that a project is not notifiable or will not require the manpower set out in regulation 3(2)(b) a client should appoint a planning supervisor to advise accordingly, prior to the absolute need to make such an appointment. For example, there may be projects where the actual number of days the persons are engaged on site may be less than 30 days, but if there are technical or other reasons for non-continuous working, such that the construction phase defined in regulation 2(1) as:

> *the period of time starting when construction work in any project starts and ending when a construction work in that project is completed*

is longer than 30 days, the exemption from regulations 4 to 12 and 14 to 19 is not available. Ultimately, if there is any doubt about the duration of the work a notice in accordance with regulation 7 should be submitted to the Executive.

Regulation 3(3) provides:

> *These Regulations shall apply to and in relation to construction work which is the demolition or dismantling of a structure notwithstanding paragraph (2).*

Due to the high risk of personal injury during demolition, any work associated with the demolition or dismantling of a structure is not included in any of the exceptions to the Regulations and therefore the Regulations apply fully.

Regulation 3(4) provides:

> *These Regulations shall not apply to or in relation to construction work in respect of which the local authority within the meaning of regulation 2(1) of the Health and Safety (Enforcing Authority) Regulations 1989 is the enforcing authority.*

The premises associated with certain activities which come under the responsibility of a local authority are set out in Schedule 1 of the Health and Safety (Enforcing Authority) Regulations 1989 and reproduced in Appendix 2.

The construction works for which a local authority is the enforcing authority are of a minor nature and extremely limited and include construction work carried out by persons who normally work on the premises and:

1.   the work is not notifiable (regulation 7); or

2.   the work is entirely internal; or

3.   the construction work is carried out in an area which is not physically segregated, the activities normally carried out on the premises have not been suspended.

Regulation 3(5) provides:

> *Regulation 14(b) shall not apply to projects in which no more than one designer is involved.*

In circumstances where there is only one designer the need for a planning supervisor to ensure co-operation with other designers simply does not arise.

Regulation 3(6) provides:

> *Regulation 16(1)(a) shall not apply to projects in which no more than one contractor is involved.*

If only one contractor is involved, in addition to the principal contractor, the need for the principal contractor to ensure co-operation between all contractors does not arise.

Regulation 3(7) provides:

> *Where construction work is carried out or managed in-house or a design is prepared in-house, then, for the purposes of paragraphs (5) and (6), each part of the undertaking of the employer shall be treated as a person and shall be counted as a designer or, as the case may be, contractor, accordingly.*

The explanation of 'in-house' is provided by regulation 2(3) as follows:

For the purposes of these Regulations –

(a)   *a project is carried out in-house where an employer arranges for the project to be carried out by an employee of his who acts, or by a group of employees who act, in either case, in relation to such a project as a separate part of the undertaking of the employer distinct from the part for which the project is carried out; and*

(b)   *construction work is carried out or managed in-house where an employer arranges for the construction work to be carried out or managed by an employee of his who acts or by a group of employees who act, in either case, in relation to such construction work as a separate part of the undertaking of the employer distinct from the part for which the construction work is carried out or managed; and*

(c)   *a design is prepared in-house where an employer arranges for the design to be prepared by an employee of his who acts, or by a group of employees who act, in either case, in relation to such design as a separate part of the undertaking of the employer distinct from the part for which the design is prepared.*

Thus, there is no prohibition on a client from providing its own design or undertaking construction work. Note however that if there is only one contractor, that contractor should be a principal contractor (regulation 6(1)). The client cannot appoint itself as principal contractor unless it is a contractor (regulation 6(2)), as provided by regulation 2(1).

It is not necessary therefore that 'each part of the undertaking of the employer' should be a separate legal entity. Departments within a local authority, for example, will be treated for the purposes of regulation 3(7) as separate persons. If the construction or design activity is carried out by 'a separate part of the undertaking', distinct from other parts of the employer, paragraphs (5) and (6) of regulation 3 will apply when the employer arranges for only one designer or contractor to supplement his own in-house resources.

Regulation 3(8) provides:

Except where regulation 5 applies, regulations 4, 6, 8 to 12 and 14 to 19 shall not apply to or in relation to construction work included or intended to be included in a project carried out for a domestic client.

The explanatory note to the Regulations explains that the Regulations do not give effect to the Temporary or Mobile Worksites Directive where construction work is carried out for a domestic client, except where regulation 5 applies with regard to a developer. The requirements of notification in accordance with regulation 7 still apply although it is the responsibility of the contractor employed by the domestic client to give such notice; for further details see Chapter 5 on the client.

Figure 4 provides a check as to whether or not the Regulations apply to a project.

Fixed term contracts

The Regulations do not apply to contract arrangements which may involve maintenance or emergency work on a frequent or irregular basis over an extended period of time. However, if there is a separate and identifiable item or section of work under the contract then such work will fall within the application of regulation 3. For this reason, unless there is only a remote likelihood that an item or section of work will fall within the scope of regulation 3, it would be good practice to appoint a planning supervisor and have a generic health and safety plan at the outset of the fixed term contract. In the event any item or section of work is likely to fall within the scope of regulation 3 the planning supervisor can advise the client and review the generic health and safety plan to ensure it is appropriate to the item or section of work to be carried out.

Extension outside Great Britain

The Regulations apply outside Great Britain to and in relation to certain premises and activities as provided by regulation 20 as follows:

> *These Regulations shall apply to any activity to which sections 1 to 59 and 80 to 82 of the Health and Safety at Work etc. Act 1974 apply by virtue of article 7 of the Health and Safety at Work etc. Act 1974 (Application outside Great Britain) Order 1989 other than activities specified in sub-paragraphs (b), (c) and (d) of that article as they apply to any such activity in Great Britain.*

Great Britain includes England, Wales and Scotland.

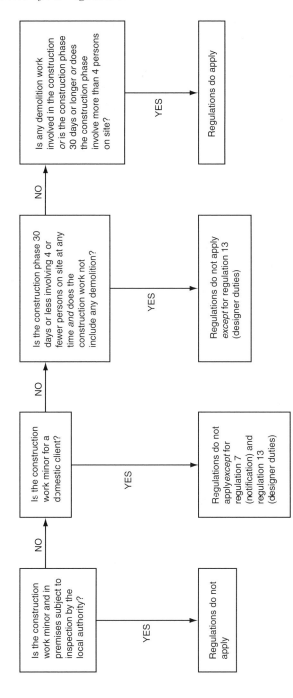

Figure 4 A check to ascertain whether and to what extent the Regulations apply

Article 7 of the Health and Safety at Work etc. Act 1974 (Application outside Great Britain) Order 1989 applies to the following activities or preparation for such activities within territorial waters: construction, reconstruction, alteration, repair, maintenance, cleaning, demolition and dismantling of any building or other structure not being a vessel.

4 Competence and resources

Competence
What is competence?
Competence under the Regulations
Allocation of resources

Competence

One of the restrictions the Regulations place on the client is the need for the client to be reasonably satisfied as to the competence of the planning supervisor or principal contractor before they are appointed.

The Regulations also require that persons appointing to the roles of designer and contractor shall be satisfied as to their competence. Paragraphs (1) to (3) of regulation 8 set out the obligations on a client, or any other person, responsible for arranging for services to be provided by one of the key roles, as follows:

(1) *No client shall appoint any person as planning supervisor in respect of a project unless the client is reasonably satisfied that the person he intends to appoint has the competence to perform the functions of planning supervisor under these Regulations in respect of that project.*

(2) *No person shall arrange for a designer to prepare a design unless he is reasonably satisfied that the designer has the competence to prepare that design.*

(3) *No person shall arrange for a contractor to carry out or manage construction work unless he is reasonably satisfied that the contractor has the competence to carry out or, as the case may be, manage that construction work.*

The assessment of competence is one of the essential features of the Regulations although recognising competence can be far from straightforward. Some assistance in determining competence is provided by paragraph (4) of regulation 8, which narrows the concept of competence as follows:

> *Any reference in this regulation to a person having competence shall extend only to his competence –*
>
> *(a) to perform any requirement; and*
>
> *(b) to conduct his undertaking without contravening any prohibition,*
>
> *imposed on him by or under any of the relevant statutory provisions.*

It will not be sufficient for persons to point to academic qualifications concerned solely with expertise in the person's chosen discipline. Of first and vital importance to 'qualify' as competent will be a knowledge of the Regulations and other 'relevant statutory provisions' which will establish the necessary awareness of the 'requirements' and 'prohibitions' imposed on competent persons.

What is competence?

The Regulations are not the only example of Parliament's intention to ensure that persons are appropriately qualified to fulfil certain statutory tasks and take responsibility for statutory obligations. A brief review of competence as applied to other regulations assists in helping to identify the essential characteristics to be found in competent persons.

The Waste Management Licensing Regulations 1994 refer to 'technical competence', which is defined as:

> *...the holder of one of the certificates awarded by the Waste Management Industry Training and Advisory Board...as being a relevant certificate of technical competence...*

The clarity and certainty introduced by the holding of a relevant certificate awarded by an identified organisation in recognising competence under the Waste Management Regulations is not shared by the Regulations or other statutory provisions, which maintain an enigmatic vagueness as to what does, or does not, constitute competence. The Electricity at Work Regulations 1989 refer to a 'competent person' and go on to stipulate that such a person should have a technical knowledge or experience in order to avoid danger when undertaking certain tasks or be under such degree of supervision as may be appropriate having regard to the nature of the work. The Factories Act 1961 had referred to a 'competent person' for the inspection and testing of hoists and lifts but without any indication whatsoever as to what necessary skills or qualifications are required. The requirement for a competent person has been carried through into the Lifting Operations and Lifting Equipment Regulations 1998. To some extent, therefore, the construction industry has learned to live with a concept of competence in certain circumstances without the need for any precise definitions or requirements.

Decisions of the courts provide some guidance to understand the notion of competence. Competence as referred to in the Construction (Working Places) Regulations 1966 is applied to persons responsible for erecting, dismantling and installing scaffolds. A competent person is required to be responsible for inspection and supervision of competent workmen, 'possessing adequate experience of such work'. In the case of *Maloney -v- A Cameron Ltd 1961 2 All ER 934*, the Court of Appeal were prepared, without further comment, to recognise the competence of three painters for the purposes of erecting trestles in accordance with the Building (Safety, Health and Welfare) Regulations 1948.

The Quarries Order 1956 requires the manager of a quarry to make and ensure the efficient carrying out of arrangements to secure that every inspection is carried out or done by a 'competent person'. In the absence of any definition of 'competent' Winn J in *Brazier -v- Skipton Rock Co. Ltd 1962* said:

> I am not prepared to hold either that 'competent' means the most competent person available to the owners of the quarry or the manager, or that it means that he shall be so competent that he never makes a mistake. In my judgment, it means a man who, on a fair assessment of the requirements of the task, the factors involved, the problems to be studied and the degree of risk danger implicit, can fairly, as well as reasonably, be regarded by the manager, and in fact is regarded at the time by the manager, as competent to perform such an inspection.

The judge went on to say:

> ...experience is so often of much greater value than book-learning.

A later decision by Cantley J in *Gibson -v- Skibs A/S Marina 1966* in relation to the Dock Regulations 1934 elaborated:

Who is 'a competent person' for the purpose of such an inspection? This phrase is not defined. I think that it is obviously to be taken to have its ordinary meaning of a person who is competent for the task. I think that a competent person for this task is a person who is a practical and reasonable man, who knows what to look for and knows how to recognise it when he sees it.

Competence is not demonstrated simply by academic qualifications. A well-qualified heating and ventilating engineer might know how to set up and record the measurements from a smoke test but that is not necessarily evidence of competence unless the engineer knows where to set up the test and what the significance of the results mean for the next step in the decision making process.

The Framework Regulations 1999, which implements the Framework Directive, at regulation 6(5) provides:

A person shall be regarded as competent for the purposes of paragraphs (1) and (8) where he has sufficient training and experience or knowledge and other qualities to enable him properly to assist in undertaking the measures referred to in paragraph (1).

The relevant ACOP provides some guidance to employers who are obliged to appoint competent persons to assist them in the fulfilment of the employer's obligations under the Framework Regulations. The ACOP draws attention to the fact that competence, as it applies to the Framework Regulations, does not require the possession of any particular skills or qualifications. In the case of simple situations, the Guidance at paragraph 51 of the ACOP suggests that competence may be satisfied by the following:

(a) *an understanding of relevant current best practice;*

(b) *an awareness of the limitations of one's own experience and knowledge; and*

(c) *the willingness and ability to supplement existing experience and knowledge, when necessary by obtaining external help and advice.*

The need for such basic requirements does not absolve the employer from taking into account the need for a competent person to demonstrate a knowledge and understanding of the work involved. The Guidance recognises that the more complex or highly technical a situation becomes, only appropriately qualified specialists possessing specific knowledge and

skills will be competent. The Guidance promotes membership of a professional body or similar organisation at an appropriate level and in an appropriate area of expertise of health and safety as demonstrating competence.
Alternatively, the Guidance suggests that competence-based qualifications
accredited by the Qualifications and Curriculum Authority and the
Scottish Qualifications Authority may provide a guide as to competence.

Competence under the Regulations

Regulation 8 requires that a client appointing a planning supervisor or
principal contractor, or a person appointing a designer or contractor shall
be 'reasonably satisfied' as to the competence of the appointee before the
appointment.

The extent to which any person requiring to be reasonably satisfied
as to another's competence is referred to in regulation 2(5)(a) which
provides:

> Any reference in these Regulations to a person being reasonably
> satisfied –
>
> (a) *as to another person's competence is a reference to that
> person being satisfied after the taking of such steps as it is
> reasonable for that person to take (including making reason-
> able enquiries or seeking advice where necessary) to satisfy
> himself as to such competence...*

Inevitably, projects of a straightforward nature will require less extensive checks or enquiries as to competence compared with projects of
greater value and complexity. In many cases a client will have little experience or understanding of the construction industry and will need the
assistance of professional advice from a third party before making an
appointment. The planning supervisor is probably the best source of
advice for all appointments, except that of himself. He is required to provide such advice on competence by virtue of regulation 14(c)(i), but this
begs the question as to what extent, in practical terms, should a client be
satisfied of the competence of third-party advisers to advise on the
appointment of a planning supervisor? Whatever the answer, the responsibility to be reasonably satisfied of another's competence before making an
appointment cannot be delegated or shared.

Demonstrating compliance with 'reasonable satisfaction' as to the
competence of the intended appointee should ideally be the outcome of a
systematic and documented appraisal. The detail and scope to which the

appraisal should extend will depend upon each particular project and will reflect the scale and complexity of the tasks to be performed. The appointer may rely on published information, reputation and promotional literature at the lower end of the scale, whereas for projects of greater scale and complexity specific enquiries are likely to be more appropriate. There have been successful prosecutions by the Executive on the grounds that there was no assessment of competence.

Allocation of resources

Regulation 9, under the heading, 'Provision for Health and Safety' provides:

> *(1)   No client shall appoint any person as planning supervisor in respect of a project unless the client is reasonably satisfied that the person he intends to appoint has allocated or, as appropriate, will allocate adequate resources to enable him to perform the functions of planning supervisor under these Regulations in respect of that project.*

> *(2)   No person shall arrange for a designer to prepare a design unless he is reasonably satisfied that the designer has allocated or, as appropriate, will allocate adequate resources to enable the designer to comply with regulation 13.*

> *(3)   No person shall arrange for a contractor to carry out or manage construction work unless he is reasonably satisfied that the contractor has allocated or, as appropriate, will allocate adequate resources to enable the contractor to comply with the requirements and prohibitions imposed on him by or under the relevant statutory provisions.*

At the first consultation stage the draft Regulations went beyond the Temporary or Mobile Worksites Directive by requiring a client to ensure that adequate financial provision for compliance with all relevant health and safety statutory provisions was made for a project. It was the Commission's view at that time, that the main health and safety costs of a project should be capable of assessment and included in tenders. The requirement on a client, to ensure adequate financial provision had been made for a project, disappeared from subsequent drafts and does not appear in the Regulations. Nonetheless, a client or planning supervisor may require information on the allocation of financial resources to matters identified specifically in the health and safety plan.

As it is, all parties to a project, except the client, are required to allocate or intend to allocate adequate resources to ensure compliance with the Regulations and, in the case of contractors, other relevant statutory provisions.

If the respective appointers are satisfied that adequate resources have been, or will be, allocated to the project by the prospective appointees, and that those parties have costed their respective contributions, the cost of the project must *ipso facto* include an adequate allocation of financial resources for the appropriate health and safety requirements. Note that significant financial resources may be required by a client to comply with the Regulations with particular regard to assessing competence and the allocation of resources by others. Despite the fact that there is no express requirement for the client to provide or allocate adequate resources it is unlikely that full compliance with the Regulations will be possible without such resources.

The original draft of the Regulations also referred expressly to the allocation of time to a project as a fundamental resource to comply with all the relevant statutory provisions with regard to health and safety. Unfortunately the emphasis on time as a resource also disappeared from the Regulations. The amount of time with which an appointer should be reasonably satisfied will depend on the planned durations for design and construction and the scale and complexity of the project.

The extent to which any person requiring to be reasonably satisfied as to the allocation of resources by another person is set out in regulation 2(5)(b), which provides:

> *Any reference in these Regulations to a person being reasonably satisfied* –
>
> (b)   *as to whether another person has allocated or will allocate adequate resources is a reference to that person being satisfied that after the taking of such steps as it is reasonable for that person to take (including making reasonable enquiries or seeking advice where necessary)* –
>
> (i)   *to ascertain what resources have been or are intended to be so allocated; and*
>
> (ii)   *to establish whether the resources so allocated or intended to be allocated are adequate.*

The assessment of adequacy of resources is to presume some prior knowledge of the size and scope of the task in the contemplation of the appointee. For clients who are unfamiliar with the construction industry it would be prudent to obtain advice to understand the nature of the tasks to be performed, from the planning supervisor for example, or from an independent third party professional, in similar terms to the task of assessing competence.

The nature and scale of the resources to be provided by the parties to a project for any particular role will, not unlike competence, vary considerably according to its scale and complexity. To demonstrate that the appointer is reasonably satisfied a systematic and documented appraisal is recommended, as with the assessment of competence.

As to whether the resources will be adequate, the person responsible for their allocation should be prepared to demonstrate adequacy based on past experience, published data, suppliers' information and most importantly, the rationale and calculations to fix staffing and distribution of health and safety management tasks.

There has been at least one prosecution that referred to lack of competence and proper allocation of resources involving a contractor. Unfortunately, the decision of the court was not fully reported and the matters that were taken into account are not therefore known.

5   The client

Introduction

During the first six years that the Regulations have been in force, organisations and individuals deemed to be performing the role of client have been prosecuted more often when compared with the total of the prosecutions for all the other roles.

The role of the client is central to the procurement process and, therefore, the Regulations recognise the important influence of the client in creating the best environment to promote the practice of good health and safety management. In a recent survey, 63% of the respondents said that clients did not understand their safety responsibilities and 75% were of the view that health and safety competence was not given sufficient priority when awarding contracts. Unless there is a significant improvement in the client's procurement procedures to take account of health and safety, clients will continue to be at the top of the CDM prosecution league.

Definition

The client is defined in regulation 2(1) which provides:

> *'Client' means any person for whom a project is carried out,*
> *whether it is carried out by another person or is carried out*
> *in-house.*

By virtue of The Interpretation Act 1978, 'person' includes corporations and limited companies as well as individuals. Therefore, anybody or any organisation embarking on a project, which includes or is intended to include construction work, is subject to the Regulations. However, identifying the client may not be without some confusion as between landowners, investment purchasers, funders, tenants or developers. In circumstances where it may be unclear as to the identity of a client out of several possible choices, a 'lead' client should be appointed. Regulation 4 provides for one of the clients to make a declaration where that client is to act as the only client: see later under 'Clients and Agents of Clients'. Some special types of clients are considered below.

Regulation 2(1) includes within the definition of clients those for whom a project is 'carried out in-house'. The definition of 'in-house' related to the carrying out of a project is set out in regulation 2(3)(a) which provides:

> *a project is carried out in-house where an employer arranges for*
> *the project to be carried out by an employee of his who acts, or by*
> *a group of employees who act, in either case, in relation to such a*
> *project as a separate part of the undertaking of the employer*
> *distinct from the part for which the project is carried out.*

Note that the employees have only to be engaged in 'a separate part of the undertaking of the employer' to be treated as in-house. There is no requirement that the employees have to be in a separate legal entity. A separate part of the undertaking may include different internal departments and units.

The Crown as client

Her Majesty's government acts in the name of the Crown, which includes all government departments and H.M. Forces.

By virtue of section 48 of HSWA 1974, the Crown is subject to the statutory duties set out in the Act, although it cannot be the subject of enforcement action by the Executive (sections 21–25), whether by way of improvement or prohibition notices or by way of prosecution (sections 33–42). Crown censures are formal recordings of a decision by the Executive that, but for Crown immunity, the evidence of a Crown body's failure to comply with health and safety law would have been sufficient to provide a realistic prospect of conviction in the courts. Accordingly, those government departments and organisations which come within the definition of the Crown have a duty to observe the provisions of the Act and the regulations made under it, including the Regulations. This is as it should be, since the declaration by Construction Minister, Viscount Ullswater in 1994, in response to the publication of the Latham Report, that:

> *The government aspires to being a best practice client, and I am happy to make that declaration. (Department of Environment press release 25th July 1994.)*

Employees of the Crown can, however, be prosecuted by virtue of section 48(2) of the Act, which is probably a sufficient incentive to ensure compliance with health and safety legislation.

In common with many other clients, the Crown is likely at times to fulfil other roles identified under the Regulations, in which case the same immunity from prosecution for the Crown and personal criminal liability of its employees will apply.

Local authorities as client

Local authorities have no immunity from prosecution like the Crown and the government departments which act in its name.

Key responsibilities

To say that without a client there will not be a project is a truism, although it serves to illustrate the importance of the client in the Regulations. Once a client has taken the decision to commission a project the initiative lies with the client to apply the Regulations.

The philosophy of improving health and safety management through the Regulations starts with the obligation to establish a professional team that will have the competence and resources to manage the project without undue risk to health and safety. The appointment of a planning supervisor is central to a client's responsibilities. The professional team must include a planning supervisor. He has the lead role for ensuring that the other members of the professional team liaise and communicate within a management framework on all matters of health and safety. The principal contractor, who is appointed after the appointment of a planning supervisor, has to plan the execution of the construction works and co-ordinate the activities of other contractors to minimise risks to health and safety.

A client is obliged to take into consideration certain factors in the selection of the planning supervisor and principal contractor as set out in the Regulations; notably, the requirements of competence and adequate resources by virtue of regulations 8 and 9 respectively. Equally, if a client appoints designers or other contractors, typically through the nominated sub-contract route, then the client is subject to the same obligation, also set out in regulations 8 and 9, as any other person appointing such persons. For a more detailed discussion on competence and resources see Chapter 4.

A client's obligations do not stop with the requirements to appoint a planning supervisor and principal contractor. The provision of information to the planning supervisor and custody of the health and safety file and the decision to proceed with the construction phase are also subject to the requirements of the Regulations.

The recording by a client of each step and action in fulfilling the requirements under the Regulations will be valuable proof to demonstrate compliance with them in the defence of a criminal prosecution. Recording the assessments and other steps required under the Regulations could be accommodated easily into a client's quality assurance procedures or form the basis for introducing quality assurance for project management.

The remainder of the chapter deals with the specific requirements imposed on a client under the Regulations.

Appointment of a planning supervisor
General

Regulation 6(1)(a) provides:

> *Subject to paragraph (6)(b), every client shall appoint –*

> *(a)   a planning supervisor...in respect of each project.*

Except where a project is not notifiable and so small that no more than four persons would be involved in the carrying out of the construction work, in accordance with regulation 3(2), all clients (except domestic clients) must appoint a planning supervisor for each project.

Who can be a planning supervisor?
The planning supervisor is a new role introduced into the construction industry by the Regulations. Accordingly, it will take some time before the expertise required for this role can be easily identified, but this will become clearer as incumbents become practised in the role. The planning supervisor can be a company, partnership or an individual. There is no requirement for the planning supervisor to be from one professional background or another – engineers, architects, surveyors, safety professionals or others will all be eligible to fulfil the role. There is also no requirement as to any particular level of qualifications, so that technician grades and tradesmen could equally be suitable, subject to the assessment of competence which is discussed below. However, it should be recognised that the planning supervisor needs a familiarity and knowledge of the design function which would confine technicians and tradesmen to the lower end of the construction project scale in terms of value, scale and complexity.

The planning supervisor does not have to be independent of all or some of the other persons fulfilling the roles envisaged under the Regulations, although independence will ensure the role is more likely to be performed in an objective and impartial manner. The Construction Industry Council has recommended that the role of planning supervisor should be subject to a separate appointment and fee arrangement.

Regulation 6(6) provides:

Paragraph (1) does not prevent –

(a)   *the appointment of the same person as planning supervisor and as principal contractor provided that person is competent to carry out the functions under these Regulations; or*

(b)   *the appointment of the client as planning supervisor or as principal contractor or as both, provided the client is competent to perform the relevant functions under these Regulations.*

The opportunity to appoint a planning supervisor who will also fulfil the role of principal contractor may have a number of advantages to a client in the ever increasing options for construction procurement routes. It is not unreasonable to assume that management contracting, in particular, and design and build projects will operate with the merged roles of planning supervisor and principal contractor. The traditional procurement route of employing separate organisations for the professional designer and contractor is unlikely to suit the merged roles of planning supervisor and principal contractor because of the usual practice of selecting the contractor after the design stage.

The overriding requirement in these situations where a client appoints one person to fulfil the merged roles of planning supervisor and principal contractor is the requirement to be satisfied as to the competence of that person to perform both roles. Although the Regulations do not expressly say so, it must be implied that the person performing both roles shall have adequate resources for both roles in order for the client to be satisfied pursuant to regulation 9.

The perceived advantages of merging the roles of planning supervisor and principal contractor may be as follows:

1. Shorter and fewer lines of communication contributing to more effective health and safety management.

2. The planning supervisor having a potentially better understanding of the construction process.

3. Less potential for contractual disputes over the content of the health and safety plan.

4. Competitive fee structure because of the lack of two separate sets of overheads.

The perceived disadvantages may include:

1. A lack of objectivity by the planning supervisor in giving the advice to the client as a result of being too close to the construction team.

2. A perception by other contractors and designers of partial advice and less than unbiased demands.

3. A lack of clarity in the evidence of the tasks performed by the planning supervisor as distinct from the tasks performed by the principal contractor.

Regulation 6(6) goes further in muddying the clarity of accountability by providing that a client can also fulfil the role of planning supervisor or principal contractor or both. Clients which have a direct labour organisation may frequently be fulfilling a joint role.

The self appointment by a client as planning supervisor, principal contractor, or both must be subject to a verifiable assessment, capable of being audited if necessary, with respect to his own competence in relation to a particular project. The importance of demonstrating the objective assessment of competence cannot be overstated and should ideally be an integral part of the client's quality assurance system. The client should also be satisfied, with respect to the allocation of resources from within his own organisation, that it is properly resourced, or intends to be, to carry out all of the different roles to which he has appointed itself.

There is no prohibition within the Regulations on the client fulfilling the role of designer and/or contractor, although, in such circumstances where another person was responsible for appointing the client as designer and/or contractor, they too would need to be satisfied as to the *client's* competence and allocation of adequate resources to fulfil the role of designer and/or contractor.

It is not necessary for the same person who has been appointed initially as planning supervisor to be involved with a project up to and including completion. Regulation 6(5) provides:

> *The appointments mentioned in paragraph (1) shall be terminated, changed or renewed as necessary to ensure that those appointments remain filled at all times until the end of the construction phase.*

The freedom to change the planning supervisor recognises that circumstances can change during the course of a project. In the initial appointment of a planning supervisor a client may be concerned to appoint a person who has particular skills relevant to the initial stages of the project design. In adopting such a strategy a client should ensure that a replacement planning supervisor has not only the competence and the resources for the subsequent construction work but is appointed with immediate effect from the termination of the previous planning supervisor's appointment, thus providing a seamless line of responsibility. The handover from one planning supervisor to another would ideally take place over a defined period of time. At the very least the changeover should be assisted by a handover meeting and status report prepared by the outgoing planning supervisor together with the latest draft of the health and safety plan and file.

When should the planning supervisor be appointed?
Regulation 6(3) provides:

> *The planning supervisor shall be appointed as soon as is practicable after the client has such information about the project and the construction work involved in it as will enable him to comply with the requirements imposed on him by regulations 8(1) and 9(1).*

A client is obliged to appoint a planning supervisor 'as soon as is practicable', which might be interpreted as without unreasonable delay. However, there would be no requirement to make such an appointment if a client is not yet committed to the 'carrying out of construction work'. In many cases a client will have required to be convinced of the technical and economic feasibility of a project before taking the decision to proceed to construction work. The planning supervisor does not necessarily have to be appointed at the feasibility stage, although an appointment at this stage can highlight the viability of various options from a health and safety point of view. An early health and safety review may eliminate some options, thus avoiding unnecessary expenditure in continuing to develop further an unviable scheme. In any event, the information which has been obtained by a client as a result of initial surveys or studies as to feasibility must be handed to the planning supervisor in accordance with regulation 11.

Delay in appointing a planning supervisor after a decision to proceed to the construction phase will not only be contrary to the Regulations but could involve uneconomic working and abortive costs if the design and preparation of tender documentation has to be revised or amended as a result of the planning supervisor's directions.

Is the planning supervisor competent?

Regulation 8(1) provides:

> *No client shall appoint any person as planning supervisor in respect of a project unless the client is reasonably satisfied that the person he intends to appoint has the competence to perform the functions of planning supervisor under these Regulations in respect of that project.*

Unless a client is satisfied that the person is competent to fulfil the role of planning supervisor the appointment would not be in accordance with the Regulations. A client can only be satisfied, by virtue of regulation 2(5)(a):

> *after taking such steps as it is reasonable for that person to take (including making reasonable enquiries or seeking advice where necessary) to satisfy himself as to such competence.*

The reference to 'seeking advice where necessary' is a pointer to the expectation that a client without in-house construction expertise should seek the advice of an independent source of construction information. In the case where a client, unaware of the Regulations, appoints a designer as the first step in a project, the designer has a duty, as set out in regulation 13(1), to ensure that the client is aware of the duties to which the client is subject.

Just because a person is judged to be competent by their peers in respect of the tasks they perform in the field of design or construction, it does not necessarily mean they would be competent within the definition of competence in the Regulations. The discussion of competence and its elusive qualities dealt with in Chapter 4 demonstrates how difficult it will be to prove or disprove competence. Whatever tests are used to assess the competence of a planning supervisor, the starting point is the understanding of the 'requirements' and 'prohibitions' referred to in regulation 8(4).

Any person who holds themselves out as competent to fulfil the role of planning supervisor must have a complete understanding of the requirements and prohibitions created by the Regulations and, 'any of the relevant statutory provisions'. It is less likely that a person will appreciate the full implications of the requirements and prohibitions without having undertaken a particular study of health and safety legislation as applied to the construction industry. A knowledge of the health and safety legislation relevant to the construction industry can be evidenced by specialist qualifications obtained by examination in occupational health and safety or assessment and attendance on recognised courses. Training is only one aspect of demonstrating competence, as highlighted in the detailed discussion on competence in Chapter 4.

The tasks of a planning supervisor make it imperative that they have relevant experience of design and appreciate the dynamics of the design process. This is particularly important for the planning supervisor's tasks of reviewing the design to ensure it includes the design considerations and information referred to in regulation 13(2) and ensuring co-operation between designers. As an example, projects which have a high electrical engineering design input would indicate that an electrical engineer would be the most appropriate background for the planning supervisor. However, the electrical engineer would be unlikely to have the competence to fulfil the planning supervisor's role in respect of building or heavy civil engineering works which might also be required. In the case of multi-disciplinary projects the client would be advised to consider the appointment of a firm or company which would be able to call upon specialists to contribute to the overall role of the planning supervisor. Consultants practising on their own account may be satisfactory for all but those projects which are complex with regard to engineering and trades content, in which case a larger organisation able to call upon a diverse range of skills would be advisable.

The importance of understanding the design role and design process should not detract from the equally important understanding of the construction process. Without a knowledge of construction techniques, methods, programming of operations and temporary works design, the planning supervisor is unlikely to be able to advise, 'any client and any contractor with a view to enabling each of them to comply with regulations 8(2) and 9(2)', as required by regulation 14(c). For this reason it is unlikely a planning supervisor will have the competence required without a significant period of experience on site, either working for a contractor or in a supervisory role.

In assessing the design and construction experience of a person prior to making an appointment as planning supervisor, a client should be satisfied that the experience matches the size, type and value of the project. A planning supervisor for a power station project will need differing expertise and experience, for instance, than might be adequate for a traditional building project.

The final aspect of competence, in considering the appointment of a planning supervisor, and the most difficult to assess is the personality of the planning supervisor, if only because it is incapable of quantitative measurement. To be successful in the role of planning supervisor a person will require abundant skills of diplomacy, persuasion and firmness of resolve.

The prudent client should approach the appointment of a planning supervisor in the same manner as the appointment of a senior management employee. The first step is to consider the scope of the task; the second step is to prepare a job profile and; finally, by a combination of interviews, references and other documentary evidence, match the candidates for the role of planning supervisor with the job profile and required qualities established during the second step.

Some of the enquiries which would comprise an assessment of competence are set out in Chapter 7 on the planning supervisor.

Adequate resources

A client who has reasonably satisfied himself as to the competence of a prospective planning supervisor is required to undertake one further assessment before making a decision on the final appointment in accordance with regulation 9(1), which provides:

> *No client shall appoint any person as planning supervisor in respect of the project unless the client is reasonably satisfied that the person he intends to appoint has allocated or, as appropriate,*

> *will allocate adequate resources to enable him to perform the func-*
> *tions of planning supervisor under these Regulations in respect of*
> *that project.*

Regulation 9(1) does not assist a client in ascertaining what would be adequate resources. Simplistically, it is not unreasonable to assume that the larger the project, the larger will be the requirement for resources that the planning supervisor has to have allocated or intends to allocate.

The resources which will be required of a planning supervisor include skilled personnel, particularly for the role of reviewing the design to satisfy regulation 14(a), and time. Time is a vital resource and a client should ensure that a planning supervisor has allocated sufficient time and man hours to the key tasks. The amount of time required is a function of the length of the design period, the number and location of designers with activities requiring co-ordination and the contractual arrangements for procuring the design.

The optimum and most effective allocation of resources requires an effective management system which can be used to monitor the correct allocation of people and other resources in the way agreed at the time when those matters were being finalised.

The management systems will include reporting and monitoring strategies, lines of communication, proposed frequency of meetings, and documentary or disk storage of data. Since different organisations will adopt different management systems this is a matter on which the client will be obliged to make a comparative judgment.

Projects which involve complex concurrent operations, multi-disciplinary design or state-of-the-art technology will produce vast amounts of information. The planning supervisor will require information technology systems to keep abreast of and monitor all the information in order to comply with the requirements of the Regulations, for all but the most simple and straightforward projects. Therefore, the information technology available to the planning supervisor will be one of the resources a client should consider before making an appointment.

The question of resources is likely to have the single biggest influence over the short-listing of possible candidates for the role of planning supervisor in the first instance. The self-employed consultant and smaller organisations are unlikely to have the optimum number of personnel and blend of necessary skills for larger projects.

For a client to be reasonably satisfied that adequate resources have been or are intended to be allocated to the project regulation 2(5)(b) requires:

> *...the taking of such steps as it is reasonable for a person in his position to take (including making reasonable enquiries or seeking advice where necessary) –*
>
> *(i)   to ascertain what resources have been or are intended to be so allocated; and*
>
> *(ii)   to establish whether the resources so allocated or intended to be allocated are adequate.*

As with the steps to be taken by a client pursuant to regulation 8(1) to ascertain competence, the importance of documenting the decision making process cannot be over stressed. A non-exhaustive checklist of those matters which a prudent client should consider is set out in Chapter 7 on the planning supervisor.

Provision of information
Regulation 11(1) provides:

> *Every client shall ensure that the planning supervisor for any project carried out for the client is provided (as soon as is reasonably practicable but in any event before the commencement of the work to which the information relates) with all information mentioned in paragraph (2) about the state or condition any premises at or on which construction work included or intended to be included in the project is or is intended to be carried out.*

The Regulations presuppose that the planning supervisor will be unable to perform his duties without the information provided by the client. However, a planning supervisor will often be in a better position to acquire information than a client. Before the type and nature of information is discussed, regulation 11(2) assists in defining the scope and extent of the information which is required to be provided by a client as follows:

> *The information required to be provided by paragraph (1) is information which is relevant to the functions of the planning supervisor under these Regulations and which the client has or could ascertain by making enquiries which it is reasonable for a person in his position to make.*

Therefore, the information to be provided by a client has to be all the information relevant to functions of the planning supervisor which is in the knowledge of the client or could be acquired by making enquiries which would be reasonable to make under the circumstances.

It is noteworthy that there is no requirement on a client to undertake any site investigation whatsoever. It is doubtful whether a client would undertake a desk study like that which would be expected of a designer to identify the geological and ground conditions prior to designing a site investigation. Enquiries which it would be reasonable for a client to make might include ascertaining the nature of operations carried out by owners of adjoining premises and enquiries of the local authority technical services department or highways authority.

It will be difficult for many clients to assess what information may or may not be relevant to the functions of the planning supervisor especially if the client has no in-house construction expertise. The Regulations refer to 'all information mentioned in paragraph (2)', and it would be safer to provide information on matters which may be irrelevant to the functions of the planning supervisor in those instances where otherwise it would be impossible to judge what is or is not relevant. In any event, there will be a number of items of information which should be in the client's knowledge and provided immediately on the appointment of the planning supervisor. This information should include:

- Address of the existing premises (if known).

- The description of the premises including maps or plans of the site.

- Description of the existing structures or plant.

- Planning history of the premises.

- Current planning conditions applicable to the premises.

- Names and identity of any tenants or users of the premises.

- Copies of any feasibility reports, surveyors or other reports prepared by organisations for and on behalf of the client in connection with the premises or the project.

- A history of past industrial usage.

- Notice of any communication with the Executive in respect of the premises.

- Notice of any prosecutions or litigation concerning the nature of the premises or adjacent premises.

- Future requirements of the client at the premises.

- General description of the project.

- Client's budget for the project.

- Existing health and safety file.

The information should be provided as soon as reasonably practicable. The regulation states that the information has to be provided before the start of the 'work' to which the information relates. Unfortunately 'work' is not defined and is probably to be interpreted as including design work otherwise the expression would have been referred to as 'construction work'. The reason for this prohibition which has the potential to delay the design phase is unclear. If the information is not supplied by the client to the planning supervisor before 'work' commences, surely it is at the commercial risk of the client that such work may be abortive.

The planning supervisor can also be placed in a difficult position by the requirement to be in possession of the information before work can commence. Since the type or scope of information referred to in regulation 11(2) is not particularised, how is the planning supervisor to know when he has all the information? The simple answer is when he is told by the client that there is no more information. However, the planning supervisor may know of the existence of information in the client's knowledge or possession or which can be ascertained by making reasonable enquiries. In such circumstances the planning supervisor may be justified in refraining from commencing any work until receiving the further information or being satisfied that the information can only be obtained by taking measures which would be unreasonable for the client, at that stage.

Typically, site investigation data on soil conditions will be very important, especially on sites where it is suspected the soil is contaminated by hazardous substances. The planning supervisor could either expect to receive this information from the client or ensure the collection of the necessary information is part of the design process, which might entail the procurement of a site investigation.

In any event, late information or changes to the information supplied in the first place may lead to late changes and abortive work to the design or preparation of contract documents and finally, delay to the construction phase. Note that the construction phase cannot start before the health and safety plan is complete and this would rely on information having been provided by the client.

In circumstances where the client is acting by an agent with expertise in the construction industry, the extent and detail of enquiries to compile the information for the planning supervisor would go beyond the readily obtainable information that a lay client could provide.

Appointment of the principal contractor
Definition
The principal contractor is defined in regulation 6(1)(b), which provides:

> *Subject to paragraph (6)(b), every client shall appoint –*
>
> *(b)   a principal contractor, in respect of each project.*

Except where a project is not notifiable and so small that no more than four persons would be involved in carrying out the construction work, in accordance with regulation 3(2), all clients (except domestic clients) must appoint a principal contractor for every project.

Regulation 6(4) provides:

> *The principal contractor shall be appointed as soon as is practicable after the client has such information about the project and the construction work involved in it as will enable the client to comply with the requirements imposed on him by regulations 8(3) and 9(3) when making an arrangement with a contractor to manage construction work where such arrangement consists of the appointment of the principal contractor.*

In many projects the appointment of the planning supervisor and designers will precede the appointment of the principal contractor by a period of time to enable the completion of the specification and design of the construction work and contract documentation. There is no reason why the principal contractor cannot be appointed much sooner, particularly if the planning supervisor is also the principal contractor, although compliance with regulations 8(3) and 9(3) will be more difficult without a health and safety plan prepared by the planning supervisor.

Which contractor?
The Regulations were prepared acknowledging the need, identified by the construction industry, that they should be adaptable to all the different routes of procuring construction projects.

In traditional contracting, under which a contractor without responsibility for design is employed by a client, it would be the natural, but not sole, choice to appoint the main contractor on being awarded the contract to the role of principal contractor.

In design and build contracting the design and build contractor may be fulfilling the role of designer and planning supervisor. There is no reason why the design and build contractor should not be appointed as principal contractor, designer and planning supervisor.

Management contractors are usually appointed for the skills and expertise which they profess to have, offering the most efficient means of co-ordinating the various work packages undertaken by works contractors. There could be distinct advantages in appointing the management contractor, if that was the selected procurement route, as planning supervisor before the design phase.

Who can be a principal contractor?
A principal contractor must be a person who is also a contractor, as defined in regulation 2(1); see Chapter 9 on Contractors.

A client has to be satisfied before an appointment that the contractor is competent and has, or intends, to allocate resources to the project in accordance with regulations 8(3) and 9(3), which provide:

> 8(3) *No person shall arrange for a contractor to carry out or manage construction work unless he is reasonably satisfied that the contractor has the competence to carry out or, as the case may be, manage, that construction work.*

and:

> 9(3) *No person shall arrange for a contractor to carry out or manage construction work unless he is reasonably satisfied that the contractor has allocated or, as appropriate, will allocate adequate resources to enable the contractor to comply with the requirements and prohibitions imposed on him by or under the relevant statutory provisions.*

At the stage of appointing the principal contractor the client will have already appointed the planning supervisor. The client will undoubtedly rely, in many cases, on the advice of the planning supervisor which he is bound to provide if asked by virtue of regulation 14(c).

When inviting tenders the client should ideally have already selected tenderers who can demonstrate competence and an intention and ability to allocate adequate resources. Even when inviting tenders under the Public Works Contracts Regulations 1991 the awarding authority is entitled to request from potential tenderers information relating to their health and safety policy and record. This is information which, although not expressly included in the criteria listed for assessing whether a contractors meets the minimum standards of technical capability, may be taken into consideration by the awarding authority when deciding whether to invite a contractor to tender (*General Building and Maintenance -v- Greenwich London Borough Council 1993* (unreported), *Building Law Monthly*, Vol. 10, Issue No. 8, August 1993).

The earlier comments in this chapter on the competence and allocation of resources by the planning supervisor apply, as appropriate, to the principal contractor. The chapters on competence and resources and the principal contractor discuss in more detail the combined issues of competence and resources with regard to contractors.

Sub-contractors
The choice of principal contractor facing a client may appear on first consideration as limited to the 'main contractor', easily recognisable in the traditional form of construction procurement and normally subject to standard terms and conditions. However, assuming a contractor has the competence and adequate resources to perform the requirements imposed on a principal contractor there is no reason why a sub-contractor could not be appointed as principal contractor at any point in the construction phase, although there are various reasons why the appointment of a person, who is not the main contractor, may not be desirable from a commercial point of view.

Nominated sub-contractors
In the case of nominated sub-contractors a client may decide the influence of the nominated sub-contract works, including specialist design if appropriate, is so significant to the health and safety management of the project that the nominated sub-contractor should be appointed as principal contractor. This may be particularly so if the sub-contractor, who it is intended will be nominated, has already entered into a contract with the client that the appointment as principal contractor could be made before that of the main contractor.

In circumstances where the proposed nominated sub-contractor has a significant influence on the construction work his early appointment as principal contractor, subject to the client's assessment of competence and adequate resources, could be beneficial in assessing main contract tenders and liaising with prospective main contractors. However, his contribution to the health and safety plan would probably be limited due to the lack of input from other contractors. Once the nomination had been perfected and the doctrine of privity of contract applied, which binds the nominated sub-contractor in a contractual relationship with the main contractor, continuation in the role of principal contractor would require a separate contract of appointment with the client.

The appointment of a nominated sub-contractor during the construction phase as principal contractor could suffer from the potential to cause commercial conflicts of interest, arising from serving two masters at the same time. Moreover, the lack of contractual pressure which could be brought to bear on the other sub-contractors of the main contractor could lead to difficulties of achieving compliance with directions of the nominated sub-contractor as principal contractor.

The drawbacks associated with a nominated sub-contractor being the principal contractor apply equally to domestic sub-contractors, not least the lack of any contractual influence over a main contractor's selection of sub-contractors. Despite the fact that any suitable contractor can be a principal contractor it is very unlikely that sub-contractors, whether nominated or not, will have the knowledge or experience of the overall works covered by the contract. The appointment of the principal contractor should be limited to main contractors unless there are significant advantages to do otherwise.

When can the construction phase start?

The construction phase is defined in regulation 2(1) as:

> *the period of time starting when construction work on any project starts and ending when construction on that project is completed.*

The planning supervisor's primary responsibility is to ensure that a health and safety plan has been prepared in accordance with regulation 15. Despite the planning supervisor's role in developing the health and safety plan, a client has the responsibility to ensure that the health and safety plan has been prepared so as to comply with regulation 15(4) before the start of the construction phase by virtue of regulation 10, which provides:

> *Every client shall ensure, so far as is reasonably practicable, that the construction phase of any project does not start unless a health and safety plan complying with regulation 15(4) has been prepared in respect of that project.*

This requirement on a client creates an important hold point. If a health and safety plan is inadequately prepared, the construction phase should not start.

A client's duty is limited to ensuring 'so far as is reasonably practicable', that the health and safety plan complies with regulation 15(4). In circumstances where a client does not have the in-house expertise, the advice of the planning supervisor or other independent adviser that the health and safety plan complies with regulation 15(4) may be sufficient. The importance of this provision is to prevent the construction phase starting whilst the health and safety plan is still in preparation. This will undoubtedly lead on occasion to postponing the commencement of site works, or delaying the date of possession of the site. Delays in preparing the health and safety plan may lead to claims for extension of time from contractors to the extent that the delay has been caused by the planning supervisor, designers or the client.

The time allowed by a client from submission of a tender to commencing site works should take full account of the development of the health and safety plan by the principal contractor.

At the least, the ACOP identifies certain aspects of the health and safety plan which should be developed in detail before the construction phase begins; these include:

- management organisation;

- site rules; and

- emergency procedures.

For certain projects, the design and planning of all the construction work will not be complete when the construction phase is planned to start. In those circumstances the ACOP recognises that a client is left to make a judgment only on those parts of the plan which have been developed for the work packages which are involved at the start of construction.

The main contract should provide that construction work can only start on site subject to the agreement of the health and safety plan in respect of all the works or the first stage of the works, and may continue to such a point but no further until the health and safety plan has been agreed in respect of the second and subsequent stages of the works. This agreement should be evidenced by receipt of a notice from the client of his satisfaction of the health and safety plan in respect of the works to which it applies. Such notice is confirmation that regulation 15(4) has been complied with to avoid any doubt or argument over the adequacy of the plan.

Custody of the health and safety file

Regulation 12(1) provides:

> Every client shall take such steps as it is reasonable for a person
> in his position to take to ensure that the information in any health
> and safety file which has been delivered to him is kept available
> for inspection by any person who may need information in the file
> for the purpose of complying with the requirements and prohibi-
> tions imposed on him by or under the relevant statutory provi-
> sions.

The planning supervisor has to ensure that a health and safety file is prepared for each structure comprised in the project. On completion of construction work for each structure the planning supervisor is required to
deliver the health and safety file to the client.

In well-managed projects clients should receive 'as-built' drawings at
the end of the project. The requirement in the Regulations to hand over the
health and safety file, which should include the 'as-built' drawings, is
therefore an extension of existing practice. There is the important difference that a client cannot simply file the health and safety file away never
to be seen, or found again, which is so commonly the case with any
'as-built' drawings.

The health and safety file may be needed by the Executive, or any other
party to the project, to obtain vital information for various purposes
including evidence for prosecution, civil litigation or subsequent extension
or renovation works and demolition. The health and safety file has to be
stored somewhere that is secure and capable of being inspected on reasonable notice. The health and safety file should ideally be treated in a
similar way to other important documents relating to premises such as
lease and mortgage documentation. The common sense of this approach is
exemplified by regulation 12(2) as amended by the Amending
Regulations, which states:

> It shall be sufficient compliance with paragraph (1) by a client who
> disposes of his entire interest in the structure if he delivers the
> health and safety file for the structure to the person who acquires
> his interest in the structure and ensures such person is aware of the
> nature and purpose of the health and safety file.

An 'interest in the structure' can mean the freehold, leasehold or licence.
The only basis on which a client can part with possession of the health
and safety file is when the entire interest in a structure is transferred. Thus,

in the situation where a freeholder grants a lease of the entire premises, there still remains 'an interest' and the health and safety file should not be handed to the tenant. Moreover, the tenant does not have any statutory right to require possession of the health and safety file, regardless of whether the lease contains a full repairing covenant. Indeed, there is no obligation on a client to deliver the health and safety file to a person acquiring its entire interest in the structure, only to make it available for inspection. In any event a prudent purchaser should seek possession of the health and safety file, or at the least, a copy of the file on acquiring an interest to a structure or building.

When the health and safety file is delivered by a client to a person acquiring his entire interest in the structure, he has to ensure the recipient is aware of the 'nature and purpose of the health and safety file'. A client should bring to the recipient's attention the necessary details in writing and the health and safety file should be marked clearly for easy identification.

Clients and agents of clients

Regulation 4(1) provides:

> *A client may appoint an agent or another client to act as the only client in respect of a project and where such an appointment is made the provisions of paragraphs (2) to (5) shall apply.*

In certain circumstances, where there may be some confusion as to who the client might be for the purposes of the Regulations, it is recommended that one client is appointed to act as the only client in respect of a project. The agreement between the clients as to the identify of the 'lead' client should be part of the project documentation.

The definition of agent in regulation 2(1) is as follows:

> *'agent' in relation to any client means any person who acts as agent for a client in connection with the carrying on by the person of a trade, business or other undertaking (whether for profit or not).*

Acting as an agent in common law is difficult to define precisely. It can be characterised as a person, expressly or impliedly authorised, to act for another, who is called the principal, and who is, in consequence of, and to the extent of, the authority delegated by him, bound by the acts of his agent. For a full discussion of the nature of agency one of the standard law textbooks should be consulted.

The relationship between agents and clients

The extent of the authority of an agent will depend on the express or implied contractual relationship between the agent and the client. It is generally unwise for any party contracting with an agent, where known, to proceed without confirmation in writing of the scope of the agent's authority and responsibilities.

The terms and conditions of the appointment of an agent acting for a client or clients should not fetter the agent's powers to act as the client under the Regulations. Any person acting as an agent who found they were unable or were prevented from exercising all the duties required of a client under the Regulations would be exposed to the risk of criminal prosecution over which they had no control.

Appointment of an agent and declaration

Regulation 4(2) provides:

> *No client shall appoint any person as his agent under paragraph (1) unless the client is reasonably satisfied that the person he intends to appoint as his agent has the competence to perform the duties imposed on a client by the Regulations.*

This regulation did not appear in any of the drafts for consultation: it certainly raises some curious questions. There is no requirement that all clients should be competent, so why impose on clients the requirement to be reasonably satisfied as to an agent's competence? This requirement would be more understandable if there was an obligation on a client to consider whether they were competent to perform the duties imposed on clients by the Regulations and requiring them to appoint an agent if they concluded they were not competent.

An agent is only required to be competent to carry out the client's duties. Clients who are used to buying in all manner of services will undoubtedly apply their own commercial common sense. Relevant factors will include experience or knowledge of the client's business, perhaps evidenced by membership of appropriate trade associations or institutions and an appreciation of the Regulations, construction industry and building practice.

Who is the client able to approach to seek the advice referred to as part of the reasonable steps in regulation 2(5) to ascertain competence? Unlike the consideration of competence before appointing a principal contractor or designer the planning supervisor is not obliged to provide advice, if appointed. Similar to appointing a planning supervisor if the client has any doubts he should consult with third parties who have an understanding and appreciation of the client's duties, typically persons engaged in the role of planning supervisor on other projects.

Unlike all the other persons under the Regulations who are required to be competent and allocate adequate resources, there is no requirement on an agent to allocate adequate resources. It should not be doubted that clients will have to allocate resources to perform their duties under the Regulations, so why is there no requirement for clients to be reasonably satisfied that an agent has allocated, or intends to allocate, adequate resources? Prudent clients will take their own steps to be reasonably satisfied as to the agent's allocation of resources.

The agent or 'lead' client is required to make a declaration to the Executive in accordance with regulation 4(3), which provides:

> *Where the person appointed under paragraph (1) makes a declaration in accordance with paragraph (4), then, from the date of receipt of the declaration by the Executive, such requirements and prohibitions as are imposed by these Regulations upon a client shall apply to the person (so long as he remains so appointed as such) as if he were the only client in respect of that project.*

For a declaration to be effective the Executive must receive a declaration which complies with the requirements of regulation 4(4) which provides:

> *A declaration in accordance with this paragraph –*
>
> *(a)* *is a declaration in writing, signed by or on behalf of the person referred to in paragraph (3), to the effect that the client or agent who makes it will act as client for the purposes of these Regulations; and*
>
> *(b)* *shall include the name of the person by or on behalf of whom it is made, the address where documents may be served on that person and the address of the construction site; and*
>
> *(c)* *shall be sent to the Executive.*

A suggested form of declaration can be found in Appendix 3. In circumstances where an agent's appointment is terminated a fresh declaration should be made and sent without delay to the Executive.

The Executive is bound to acknowledge receipt of a declaration in writing to the client or agent confirming the date the declaration was received by the Executive, in accordance with regulation 4(5) which provides:

> *Where the Executive receives a declaration in accordance with paragraph (4), it shall give notice to the person by or on behalf of whom the declaration is made and the notice shall include the date the declaration was received the Executive.*

The making of a declaration is not mandatory. Regulation 4(6) takes account of the situation where no declaration is received by the Executive, which provides:

> *Where the person referred to in paragraph (3) does not make a declaration in accordance with paragraph (4), any requirement or prohibition imposed by these Regulations on a client shall also be imposed on him but only to the extent it relates to any matter within his authority.*

Without a declaration, the client or clients, and agent or agents, will share the burden of the requirements and prohibitions imposed by the Regulations, to the extent that an agent has any authority with regard to such requirements and prohibitions. In the absence of any express contractual authority given to an agent or single client such authority would be a matter of fact to be determined from the actions of the agent, or ostensible authority. Indeed, an agent may hold himself out as having an authority which exceeds an express contractual authority; in which case, the agent may be liable for those matters which to a third party, appeared to be within his authority.

Domestic clients and developers

The domestic client is defined in regulation 2(1) which provides:

> *a client for whom a project is carried out not being a project carried out in connection with the carrying on by the client of a trade, business or other undertaking (whether for profit or not).*

A domestic client is typically someone for whom a house is being built for that person's occupation, or is having an extension to his/her house or having minor building or engineering works in the garden or grounds of the house. However, the status of a domestic client can easily be lost. If a person has an extension to their house because part of the house is used for the carrying on of a trade or business such as a dental surgery or home office, the Regulations will apply provided that the project comes within the scope of the Regulations, by virtue of the number of persons employed and the length of the construction period.

The domestic client does not have to concern himself with the Regulations. Regulation 3(8) provides:

> *Except where regulation 5 applies, regulations 4, 6, 8 to 12 and 14 to 19 shall not apply to or in relation to construction work included or intended to be included in a project carried out for a domestic client*

Note that in all cases regulations 7 and 13 apply, although the contractor only has the responsibility to notify the Executive under regulation 7 if he has reasonable grounds for believing that the project is notifiable. The requirement for, and the form and timing of the notice, is set out in regulation 7, paragraphs (5) and (6) which provide:

> (5) *Where a project is carried out for a domestic client then except where regulation 5 applies, every contractor shall ensure that notice of the project is given to the Executive in accordance with paragraph (6) unless the contractor has reasonable grounds for believing that the project is not notifiable.*

> (6) *Any notice required by paragraph (5) shall –*

> (a) *be in writing or such other manner as the Executive from time to time approve in writing;*

> (b) *contain such of the particulars specified in Schedule 1 as are relevant to the project; and*

> (c) *be given before the contractor or any person at work under his control starts to carry out construction work.*

Schedule 1 and a form of notice set out in Appendix 4 containing the full particulars of Schedule 1 should be completed as far as such particulars are relevant.

The designer, who is subject at all times to the obligations under regulation 13, is not relieved of any of the obligations simply because the work is for a domestic client.

The Regulations have addressed the situation of commercial house developers who sell domestic premises to a domestic client. The developer shall, in accordance with regulation 2(1), be construed in accordance with regulation 5(1). Regulation 5 provides:

> *(1)*   *This regulation applies where the project is carried out for a domestic client and the client enters into an arrangement with a person (in this regulation called 'the developer') who carries on a trade, business or other undertaking (whether for profit or not) in connection with which –*
>
> > *(a)*   *land or an interest in land is granted or transferred to the client; and*
> >
> > *(b)*   *the developer undertakes that construction work will be carried out on the land; and*
> >
> > *(c)*   *following the construction work, the land will include premises which, as intended by the client, will be occupied as a residence.*
>
> *(2)*   *Where this regulation applies, with effect from the time the client enters into the arrangement referred to in paragraph (1), the requirements of regulations 6 and 8 to 12 shall apply to the developer as if he were the client.*

The essential requirements of the 'arrangement' are set out in paragraphs (a) to (c) of regulation 5(1). The arrangement normally will be a contract or conveyance of the land on which premises are still in the process of being constructed or are intended to be constructed. The arrangement does not have to include a sale, since an interest in the land may be no more than an option granted to a domestic client over a plot of land.

If a domestic client has no intention of occupying the premises as a residence, perhaps with a view to an investment opportunity or letting, and has the benefit of an arrangement with a developer without disclosing the true intention behind the arrangement, the domestic client will be subject to the full effect of the Regulations provided that the project falls within the scope of the Regulations. A prudent developer would be well advised to make enquiries of a domestic client to have confirmed that the premises will be occupied as a residence by the domestic client or other person, a relative for instance, without paying rent. A developer's obligations to comply with the Regulations, in the role of client, would cease on the conclusion of an arrangement with a non-domestic client.

In many instances, housing developments will be sufficiently large to require notification to the Executive and the developer will be the client and have responsibility for the appointment of the planning supervisor and principal contractor. The effect of the Regulations is to ensure that the obligation of the developer is not diminished or shared with the bona fide purchaser of premises for residential occupation.

Basic checklist of considerations for a client

1. Do the Regulations apply to the project?

Client and agent

2. Are there other persons who may be the client?

3. Who will be the client for the purposes of the Regulations?

4. If you are the client, have you made a declaration in accordance with regulation 4(4)?

5. Do you wish to appoint an agent?

6. Have you ensured that the agent is competent before making the appointment?

7. If you have appointed an agent, ensure he makes a declaration in accordance with regulation 4(4).

Planning supervisor

8. Are you reasonably satisfied as to the planning supervisor's competence and allocation of adequate resources?

9. Have you appointed a planning supervisor, or appointed yourself, in accordance with regulation 6?

10. Has the planning supervisor notified the Executive of the project in accordance with regulation 7?

11. Have you supplied the information to the planning supervisor in accordance with regulation 11?

Principal contractor

12. Have you sought the advice of the planning supervisor before arranging the appointment of a principal contractor?

13. Are you reasonably satisfied as to the principal contractor's competence and allocation of adequate resources?

14. Have you arranged for a principal contractor to carry out or manage construction work, or appointed yourself, in accordance with regulation 6?

Designer

15. Have you sought the advice of the planning supervisor before arranging the appointment of a designer?

16. Are you reasonably satisfied as to the designer's competence and allocation of adequate resources?

17. Are you a designer within the meaning of the Regulations, or have you arranged for a designer to prepare a design?

Contractor

18. Have you sought the advice of the planning supervisor before arranging the appointment of a contractor?

19. Are you reasonably satisfied as to the contractor's competence and allocation of adequate resources?

20. Have you arranged for a contractor to carry out or manage construction work, or have you appointed yourself a contractor?

Health and safety plan and file

21. Have you checked that the health and safety plan complies with regulation 15(4) before giving approval to the start of the construction phase?

22. Have you taken delivery of the health and safety file from the planning supervisor?

23. Have you stored the health and safety file so that it is available for inspection by any interested person?

6 The designer

Introduction

The role of designer, in accordance with the Regulations, is probably the most difficult to assess from an objective standpoint. Nonetheless, most people employed in the construction industry will concede that the role of designer is underestimated when considering the potential for minimising health and safety risks during the construction and maintenance phases. Yet, in the face of that instinctive conclusion, the number of prosecutions involving designers are a very small percentage of the total number of prosecutions. The chief inspector in the Executive for construction expressed the view, five years after the Regulations were introduced, that the Regulations had failed to make an impact on designers. He went on to say that the Executive would be looking for new ways to carry out inspection and enforcement so that it could target designers. At the time of writing, the Executive has instigated trial audits of volunteer design companies with the intention to devise a methodology to check designers' compliance with the Regulations in the future.

Designers can expect increasing scrutiny from the Executive and this will, no doubt, lead to more prosecutions and possibly more professional negligence claims. But, it is clear that simply exhorting designers to design structures that are more safe to build has not been enough. Designers need to be trained to think 'safety'. The role of the Executive is all too frequently after the event. Learning about 'safe design' after there has been an accident and a prosecution is not nearly as effective as introducing the health and safety discipline in the day to day work in the design office.

Definition

It is ironic, on the basis that a relatively small number of designers have been prosecuted, that it was a prosecution of a designer which was appealed to the Court of Appeal that has led to the first substantive change to the Regulations. The Construction (Design and Management) (Amendment) Regulations 2000 ('Amending Regulations') has revised the definition of designer.

The Court of Appeal held in *R. -v- Paul Wurth SA* that regulation 13(2)(a) applied only to a designer who 'prepares a design'. The case involved a fatal accident that occurred when a conveyor fell and crushed a man. The conveyor fell because of a simple fault in the design where the latching devices should have incorporated a locking or securing pin but the drawings used by the erectors had not shown the pin. Wurth had employed Fairport Engineering Limited to convert its design into manufacturing and construction drawings, and the latter had contracted the manufacture to the Universal Conveyor Company Limited. The issue addressed by the Court of Appeal was whether Wurth had 'prepared' the 'design' of the latching devices within the meaning of the words in regulation 13(2)(a). The Court found that the case could not be brought within regulation 13(2)(a) because that applied only to a designer who 'prepares a design' and Wurth had not prepared the relevant design. This loophole was remedied by the Amending Regulations. In particular, the new definition of designer is limited to a person who 'prepares a design'.

A designer is defined in regulation 2(1) as:

> *...any person who carries on a trade, business or other undertaking in connection with which he prepares a design.*

Unlike the definition of 'contractor', in this definition the designer is not expressed to be a person carrying on, 'a trade, business or other undertaking (whether for profit or not)'. If the omission of the phrase 'whether for profit or not' was not accidental the alternative is to assume the phrase has been 'designedly omitted' (*Union Bank of London -v- Ingram (1882) 20 ChD 463*). Does the omission have any relevance? The answer is probably no, because an 'undertaking' does not have to be one that makes a profit, i.e. a charity or local authority, and therefore the term 'whether for profit or not' neither adds nor takes away anything from the definition, which is the same as the definition for a designer. Suffice to say that the omission is puzzling and surprising.

A designer is a person who designs – or is it? To answer what appears to be a cryptic question an understanding of design is required. The definition of design in regulation 2(1) provides:

> *'design' in relation to any structure includes drawing, design details, specification and bills of quantities (including specification of articles or substances) in relation to the structure*

It is within most people's contemplation that design would include the preparation of drawings, calculations and specifications, but note in particular that the preparation of bills of quantities is included within the definition of design, thus bringing quantity surveyors within the category of designer. The inclusion of bills of quantities hints at the importance of the allocation of financial provision which had been omitted from the original consultative document for the Regulations. There is also no distinction between permanent works or the temporary works required for construction which will either be taken down or left within the fabric of the structure.

The list of persons who might prepare drawings, specifications or bills of quantities is almost endless. Although the designer is confined to the person who prepares the design, there is still a reference in regulation 2(2) for a person arranging for another to prepare a design. To that extent, regulation 2(2) from the designer's point of view is otiose. Regulation 2(2) is preserved for the arranging for another person to carry out or manage construction work. This is dealt with by a new regulation 2(3A), which provides:

> *...any reference in these Regulations to a person preparing a design shall include a reference to his employee or other person at work under his control preparing it for him; but nothing in this paragraph shall be taken to effect the application of paragraph 2.*

Selecting a designer

Regulation 8(2) provides:

> *No person shall arrange for a designer to prepare a design unless he is reasonably satisfied that the designer has the competence to prepare that design.*

No single person on a project is charged with the responsibility to appoint a designer. On large and complex projects there may be many different organisations appointing designers. There are only two factors to be taken into account by any person appointing a designer: these are competence and resources.

Competence

In accordance with regulation 8(2) a person appointing a designer has to be reasonably satisfied as to their competence.

The assessment of competence is not concerned solely with design ability but with the competence to 'perform any requirement' and avoid 'contravening any prohibition' in accordance with regulation 8(4). Qualifications are not the sole criteria by which it is possible to assess the competence of a designer as discussed in Chapter 4.

In the first instance, any person who is considering appointing a designer can seek advice from the planning supervisor, by virtue of regulation 14(c)(i), and may later seek his views before actually making an appointment.

The work of designers is fundamental to the nature of the construction work. The difference between choosing a structural steelwork frame in preference to a reinforced concrete frame, as an example, will have a profound impact on the site activities and health and safety measures to be implemented. A designer may be able to design the frame in steel or reinforced concrete. If the designer does not have an appreciation of the impact on health and safety aspects the final selection will not be based on a complete set of criteria. Such a designer would not be competent for the purposes of the Regulations.

Hazards associated with construction activities have to be identified before a designer can consider the best means of eliminating or reducing the corresponding risk. It is important therefore that a designer has the skills and experience to identify hazards in the first step of risk assessment discussed below under preventative design.

The enquiries, based on the ACOP, which a person might make of the designer are likely to include the following:

1. The name of the main board director or partner with responsibility for health and safety and the name, qualifications and experience of the person providing health and safety advice and/or assistance by virtue of regulation 6 of the Framework Regulations.

2. Can the designer demonstrate a good claims record against their professional indemnity policy in respect of health and safety matters (and Employers' Liability Insurance)?

3. Has the designer or any of its personnel any convictions under any health and safety legislation?

4. Does the designer have an awareness of relevant health and safety legislation and appropriate risk assessment methods?

5. What is the accident record on past projects on which the designer has been involved?

6. What references can the designer offer from past clients?

7. Is the designer able to demonstrate a familiarity with construction processes, in the circumstances of the project, and the impact of design on health and safety? Past experience on similar projects would be a sensible means of demonstrating such familiarity.

8. Is the designer able to produce a copy of his own health and safety policy as applied to the designer's staff?

9. Is the designer able to produce a copy of the health and safety policy as applied to design work carried out by the designer?

10. Is the designer able to demonstrate that the people to be employed to carry out the work possess the appropriate skills and training? Specifically included are external resources and the review of the design against the requirements of regulation 13(2).

11. What steps does the designer take to maintain competence?

Adequate resources

Regulation 9(2) provides:

> *No person shall arrange for a designer to prepare a design unless he is reasonably satisfied that the designer has allocated or, as*

*appropriate, will allocate adequate resources to enable the designer
to comply with regulation 13.*

To be reasonably satisfied an appointer must ascertain firstly what
resources have been, or are intended to be allocated, and secondly whether
such resources are adequate.

The enquiries an appointer might make, based on the ACOP, include the
following questions:

1.   How much time has the designer allowed to fulfil the various
     elements of the designer's work? In particular, how many personnel
     does the designer have allocated to the project, or intend to allocate,
     throughout the duration of the design phase?

2.   What part of the design has been, or is intended by the designer, to
     be carried out by sub-contractors together with, where appropriate,
     relevant information as to the competence of such sub-contractors?

3.   What technical facilities are available to support the designer, partic-
     ularly in the circumstances of the project? These may include:

     (i)   computer aided design facilities;

     (ii)   technical library;

     (iii) laboratory services;

     (iv)   access to research information;

     (v)   membership of professional or trade associations;

     (vi)   document/drawing storage;

4.   Is the office accommodation available to the designer adequate to
     accommodate the allocated personnel?

5.   What method will the designer use to communicate design decisions
     to ensure that the resources to be allocated are clear?

6.   How will the designer communicate the information on remaining
     risks, after complying with the duties in regulation 13(2)(a)?

The second stage as to adequacy involves a judgment as to whether the
resources allocated, or intended to be allocated, are adequate. A client or
any contractor considering the appointment of a designer, can seek the

advice of the planning supervisor. Further enquiries of the designer to ascertain adequacy might include the following questions:

1. What are the qualifications, experience and skills of each person who the designer allocates to the design phase (including personnel of sub-contractors)?

2. Are the persons allocated by the designer to the design phase able to demonstrate an awareness of the designer's policy statements on health and safety as applied to personnel and design considerations? More importantly, can the designer produce evidence that such policy statements and other rules and guidelines are observed, other than in the breach?

3. Has the designer allocated adequate time to each stage of the design process, in particular to risk assessment and review of the design to comply with the requirements of regulation 13?

4. Has the designer a system of internal and external communication and information exchange which will ensure that the design decisions are properly recorded and disseminated and that resources are allocated appropriately?

5. Has the designer a systematic approach to ensure that information on unavoidable risks, after the duties in regulation 13(2)(a) have been complied with, will be communicated to all interested parties?

6. Can the designer demonstrate the implementation and practice of quality assurance principles? Registration by one of the accreditation organisations would be a quick and efficient means of providing the confidence that the designer has a management system incorporating items (4) and (5) above, although such registration is not absolutely necessary.

Any client or contractor appointing a designer has the opportunity of seeking advice from the planning supervisor before making an appointment by virtue of regulation 14(c)(i).

The client as designer

A client is not prevented from fulfilling the role of designer subject to being able to demonstrate that he is satisfied that his design capability, in-house or otherwise, is competent and adequately resourced to fulfil the duties imposed by the Regulations. In circumstances where there is an independent planning supervisor, it would be prudent for a client to seek his views before confirming the appointment of the designer from within his own organisation.

In some circumstances a client may wish to impose particular design principles, standards or specifications on a designer to ensure consistency of design with other projects which the client is procuring, or has procured. Whilst the liability for design in such situations becomes blurred, the client will be fulfilling the role of designer under the Regulations. A client who has 'control' over the designer has to accept that he will become subject to the requirements of regulation 13(2) to an extent which will depend on the precise level of contribution to the design. It is recommended that the definition of a designer is studied carefully.

The planning supervisor as designer

The role of planning supervisor is inescapably linked with the design process. Therefore, whilst the planning supervisor is not prevented by the Regulations from fulfilling the role of designer, there is the risk that there might be a lack of independence and objectivity, implicit in regulation 14, to review the design considerations and production of adequate information.

Many design organisations offer services as the planning supervisor and there may be particular advantages to appointing the same organisation as designer and planning supervisor. However, where possible, to preserve the 'independence' and objectivity of the planning supervisor it would be preferable to have the planning supervisor either in a different office or location from the designer or working within a completely different subsidiary company within the design organisation's group holding. The Construction Industry Council recommend that the planning supervisor should be engaged under separate terms of engagement with a separate identifiable fee.

The principal contractor as designer

The Regulations have been drafted to accommodate the ever-increasing methods of procurement in the construction industry including 'design and build'. Under the forms of contract for design and build the main contractor undertakes to design the permanent and temporary works in addition to the execution of the construction phase. The exact extent of liability as to design will be subject to the contract terms. The Regulations do not prohibit a design and build contractor from fulfilling both roles and this is likely to become commonplace.

A principal contractor without any in-house design capability, whether for temporary or permanent works, will need to appoint a designer in accordance with, and subject to, regulations 8(2) and 9(2). The assessment of competence and adequacy of design resources applies equally to temporary and permanent works design procured by a principal contractor. In circumstances where the planning supervisor has not been appointed from within the principal contractor's own organisation, there will be a much earlier liaison with the planning supervisor and overlap in the development of the health and safety plan.

Requirements on the designer

The Regulations impose two distinct duties on a designer which are:

1. A duty to notify the client;

2. A duty to design in accordance with the requirements set out in the Regulations.

A duty to notify the client

Regulation 13(1) provides:

> *Except where a design is prepared in-house, no employer shall cause or permit any employee of his to prepare for him, and no self-employed person shall prepare, a design in respect of any project unless he has taken reasonable steps to ensure that the client for that project is aware of the duties to which the client is subject by virtue of these Regulations and of any practical guidance issued from time to time by the Commission with respect to the requirements of these Regulations.*

The above regulation did not appear in any of the drafts of the Regulations during the consultation phase. However, its inclusion has avoided the obvious difficulty which a client might encounter if it was unaware of the Regulations. Ignorance is no defence in the law, but the risk of many clients, not connected with the construction industry, proceeding to an advanced stage on a project, without observing its duties under the

Regulations was very real. Such circumstances would also have 'tainted' the professional members of the project team. They would have been aiding and abetting in the commission of an offence if they were aware of the client's lack of knowledge and failed to bring this to the attention of the client. The requirements of regulation 13(1) avoid any uncertainty and confusion by imposing on a designer the duty to take reasonable steps to bring to the attention of the client the duties to which a client is subject. The Amending Regulations inserted, after the word 'prepare' where it first occurs, the words 'for him'. Thus, the requirement in regulation 13(1) does not apply to an employer who supplies his employee to a designer.

The designer's obligation to take 'reasonable steps' to bring the Regulations to the attention of the client may be fulfilled by informing the client verbally at a meeting, or during a telephone conversation. Verbal notice is always difficult to prove and it is recommended that the designer should write a letter to the client bringing the Regulations to his attention and, ideally, should be referred to in any formal terms of appointment. Note that the client should have his attention drawn to 'practical guidance issued from time to time by the Commission'. The guidance will certainly include the ACOP and any guidance notes produced by the Executive and might also include references to any reports, announcements or press releases of the Commission.

The marketing opportunities presented to designers in offering the services of the planning supervisor have been widely exploited and provide a natural incentive to comply with regulation 13(1).

A duty to design in accordance with the requirements set out in the Regulations

Regulation 13 always applies to construction work regardless of whether the project is not notifiable or the number of persons at any one time will not exceed four by virtue of regulation 3(2). Regulation 13 does not even depend upon appointments made by the client nor upon the existence of a client. Whenever a designer prepares a design which he is aware will be used for the purposes of construction work, regulation 13 applies. Thus, the Regulations apply to the design of construction products manufactured for sale without any specific purchaser or project in view.

Regulation 13(2) only applies to design which the designer 'is aware will be used for the purpose of construction work'. Accordingly, designs prepared for purposes other than for construction work are not subject strictly to the Regulations, which would include feasibility studies, schemes for cost comparisons or for illustrative purposes only.

Nonetheless, the health and safety aspect of studies at the pre-project stage cannot be ignored if a true comparison of the alternatives is to be undertaken. Therefore, it may be of benefit to a client to appoint a planning supervisor to consider the health and safety aspects of such studies. Indeed, it is likely that such an early appointment would be sound commercial practice. In an obvious example the comparison of a tunnel project, which could be carried out by compressed air tunnelling or deep open cut would not be complete without an appreciation of the risks to health and safety involved in compressed air tunnelling compared with working at the bottom of deep excavations.

Preventative design

Regulation 13(2) provides:

Every designer shall –

(a) *ensure that any design he prepares and which he is aware will be used for the purposes of construction work includes among the design considerations adequate regard to the need –*

   (i) *to avoid foreseeable risks to the health and safety of any person at work carrying out construction work or cleaning work in or on the structure at any time, or of any person who may be affected by the work of such a person at work;*

   (ii) *to combat at source risks to the health and safety of any person at work carrying out construction work or cleaning work in or on the structure at any time, or of any person who may be affected by the work of such a person at work; and*

   (iii) *to give priority to measures which will protect all persons at work who may carry out construction work or cleaning work at any time and all persons who may be affected by the work of such persons at work over measures which only protect each person carrying out such work.*

The ACOP requires the designer to weigh the risk to health and safety produced by a feature of a design against the cost of excluding that feature by:

1. designing to avoid risks to health and safety;

2.   tackling the causes of risks at source; or if this is not possible,

3.   reducing and controlling the effects of risks by means aimed at pro-
      tecting anyone at work who might be affected by the risks and so
      yielding the greatest benefit.

The ACOP recognises that the cost equation is not based solely upon
financial consequences. There is a cost associated with buildability, performance specifications, environmental impact, form and style. (The
ACOP also refers to 'fitness for purpose' although this should not be con-
fused with the standard of care expected of a designer.) The difficulty in
allocating a 'cost' to such imponderables is neatly encapsulated by Sir
Alan Muir Wood:

> Achievement is only rarely measurable in a single unit such as cost, so
> success of the ultimate product will be a question of subjective judgment.
> (*The Institution as a learning society – Note on Design, Appendix D:
> September 1994*)

It cannot be denied that construction is an inherently dangerous activity
and will never be free from risk. Indeed, the ACOP acknowledges that the
overall design process should not be dominated by the need to avoid all
risks during the construction phase and maintenance. The challenge facing
designers is the ability to seek out and discover or develop other techniques
or construction methods to produce the same or similar results than a more
inherently high risk option. At each stage the designer is required to ask of
himself, 'Can this design be constructed more safely without undue detri-
ment to the other important factors including buildability, environmental
impact and aesthetics?'.

The designer is to have adequate regard, in accordance with paragraph
(2)(a)(i) of regulation 13, to the need to avoid foreseeable risks to persons
building, maintaining, repairing, cleaning and ultimately demolishing, the
structure. Whilst the requirement is for the identification of foreseeable
risks, there is no express requirement to quantify such risks.

A risk assessment carried out by a designer should follow the normal
pattern of the following steps:

1.   Identification of hazards;

2.   A qualitative or quantitative assessment (depending on the circum-
      stances) of the risks associated with each hazard.

The detail and amount of time a designer will devote to a risk assessment for a particular structure will depend upon its scale and complexity,

but the obligation to identify foreseeable risks is an onerous one. A foreseeable risk has to be associated with a foreseeable hazard. Therefore, unless the designer is able to identify all the foreseeable hazards, not all the foreseeable risks will be included in the assessment. Note that what is, or is not, foreseeable is not subject to any reasonableness criterion, which is to say that in the event of an unusual accident to a person on a project, the risk was foreseeable, however small. The lack of a reasonableness criterion places a very high burden of responsibility on designers. Although a designer is required to have identified all foreseeable risks, the only requirement on the design is to have 'adequate regard' for such risks. Therefore, it is not a requirement of a design to eliminate all foreseeable risks. Understanding the creation and existence of hazards and the appreciation of risk, emphasises the need for competent designers who can draw on their own experience and published information to comply with paragraph (2)(a)(i).

Curiously, at the pre-construction stage, the risks to the health and safety of any person carrying out construction work included in the designer's risk assessment are based on foreseeable risks, and yet the inclusion in the health and safety plan is expressed in regulation 15(3)(c) as:

> *details of risks to the health and safety of any person carrying out the construction work so far as such risks are known to the planning supervisor or are* reasonably *foreseeable. (Emphasis added.)*

The designer may find it helpful in conducting the risk assessment exercise to break the structure down into three main sources of hazard. Firstly, the designer should examine methods by which the structure might be built in the context of his design options. The need to avoid risks completely, or to tackle them at source, or by reducing or controlling their effects will impact on design decisions. Despite the need to assess the risks during the construction phase associated with the method of construction, the Regulations do not require designers to specify construction methods or to exercise a health and safety management function over contractors, as they carry out construction work. In certain projects, the method of construction may be inextricably linked to the design philosophy, in which case the contractor will be contractually obliged to comply with such method of construction set out in the contract documents.

Secondly, the designer will have responsibility for specifying substances or equipment for use during the construction work. The manufacturers of such substances and equipment are obliged, by virtue of HSWA 1974, to produce product literature identifying any risks in handling or use, which should be considered by the designer. Not all hazards can be avoided, such as alkaline burns arising from skin contact with wet concrete, but prevention is easily effected provided that the information is communicated to the relevant persons.

Finally, the designer also has an obligation to consider persons who will be engaged in construction work and cleaning work on the structure after the completion of the construction phase. Particular hazards will include, for example, falling from height during cleaning operations, electrocution or electric burns when repairing or maintaining electrical equipment etc.

Since no design is ever likely to be free of risk there will always be residual risks even after the final design solution has taken adequate regard of all foreseeable risks. The remaining risks can be eliminated or reduced by combating at source the risks to the health and safety of workers as required by virtue of paragraph (2)(a)(ii). For example, many accidents due to falls during construction and cleaning work could be reduced by the pre-planning and design of cast-in eyebolts for the attachment of harnesses. Identifying problems associated with the build-up of exhaust gases or other fumes which can be overcome by improving ventilation in confined spaces would be another example. The examples are almost endless and are beyond the scope of this book. However, the increasing awareness of the influence of design over health and safety management of projects will be improved by a growing body of published literature and statistics.

The designer is also required to ensure that among his design considerations he gives priority to measures which afford protection, or reduce the risk to the health and safety of workers generally, rather than just concentrating on the risk to the person undertaking a particular task.

Evidence of the design considerations having regard to (i) to (iii) of regulation 13(2)(a) will be subject to review by the planning supervisor, to the extent that he is required to ensure that the design has been prepared in accordance with that regulation. It is suggested therefore that the designer develops a methodical approach to recording the design considerations with reasoned outcomes.

Adequate information
Regulation 13(2)(b) provides:

> *ensure that the design includes adequate information about any aspect of the project or structure or materials (including articles or substances) which might affect the health or safety of any person at*

> *work carrying out construction work or cleaning work in or on the structure at any time or of any person who may be affected by the work of such a person at work.*

The designer should set out clearly the principles and assumptions involved in the design, and identify and describe any special requirements, which the contractor will need to know in considering the method of construction. The designer should ensure that this information is included in the health and safety plan as the means of communicating it to the principal contractor and other contractors.

The information the designer is required to include with the design is not, as mentioned previously, limited to construction work, but also includes cleaning work, defined in regulation 2(1) as:

> *the cleaning of any window or any transparent or translucent wall, ceiling or roof in or on a structure where such cleaning involves a risk of a person falling more than 2 metres.*

Thus, where a structure has, as part of its design, any foreseeable risks involved in cleaning together with the maintenance, repair, alteration and demolition work, these should be passed to the planning supervisor so that he might comply with regulation 14(d), by including the information in the health and safety file.

Co-operation with the planning supervisor and other designers
Regulation 13(2)(c) provides:

> *co-operate with the planning supervisor and with any other designer who is preparing any design in connection with the same project or structure so far as is necessary to enable each of them to comply with the requirements and prohibitions placed on him in relation to the project by or under the relevant statutory provisions.*

The Regulations do not require any particular means of co-operation, although it is likely that the planning supervisor, in taking the lead role for co-ordinating the health and safety management, will instigate a management reporting system. The requirement to co-operate with other designers should be part of the health and safety management system because it will not be possible for a designer on large projects to be aware of all the other designers who may be involved. Indeed, the regulation refers specifically to designers engaged on the project and not necessarily on the same structure.

On projects where there are numerous structures and there exists the potential for one structure to influence the health and safety considerations of another structure, the need for co-operation between the designers is vital. Especially important will be design assumptions and criteria, particularly loading conditions and transient effects such as, for example, flooding risk.

The processes of co-operation and co-ordination should therefore avoid or reduce risks arising from the interaction between different structures or the designers working on the same structure. These processes will also ensure the planning supervisor has captured all the information necessary for incorporation into the health and safety plan and health and safety file.

During the exchange of information and dialogue between designers the planning supervisor may identify shortcomings in the design with regard to health and safety or may suggest an alternative approach. In such circumstances the designers should co-operate to achieve the optimum design solution with regard to minimising or eliminating the foreseeable risks.

Limitation on design considerations

The entirety of paragraphs (2)(a) and (b) of regulation 13, dealt with in detail above, are subject to paragraph (3), which indicates the extent to which the design should be developed with regard to health and safety matters as follows:

> *Sub-paragraphs (a) and (b) of paragraph (2) shall require the design to include only the matters referred to therein to the extent that it is reasonable to expect the designer to address them at the time the design is prepared and to the extent that it is otherwise reasonably practicable to do so.*

This paragraph, which had not been included in any previous drafts of the Regulations for consultation, recognises that the design process proceeds from the general to the specific. For example, it would be wholly unreasonable for a designer at the outline stage of design to consider in detail foreseeable risks arising out of access to parts of a structure during the construction phase. Nonetheless, a designer cannot rely on the fact that a design has not proceeded to a stage where certain health and safety matters would require consideration, as any defence, because such matters should be addressed at all times 'to the extent that it is otherwise reasonably practicable to do so'. Thus, the competent designer will recognise particular features which might indicate health and safety hazards and the design should be marked to note these are hazards which have yet to be considered in further detail at a later stage in the refinement of the design.

Basic checklist of considerations for a designer

1. If there is a client, have you taken reasonable steps to make sure that the client is aware of his duties under the Regulations?

Competence and adequate resources

2. Are you competent?

3. Are you reasonably satisfied as to the competence of any designer engaged by you?

4. Have you allocated adequate resources?

5. Are you reasonably satisfied that any designer engaged by you has or intends to allocate adequate resources?

Health and safety of persons

6. Have you ensured that your design includes among the design considerations, adequate regard to the need:

   (i) to avoid foreseeable risks (both construction and cleaning works)?

   (ii) to combat at source risks to the health and safety of persons carrying out construction work or cleaning work?

   (iii) to give priority to measures which will protect all persons at work over measures which only protect each person carrying out work?

7. Have you ensured that the design includes adequate information about any aspect which might affect the health or safety of persons carrying out construction or cleaning work?

Planning supervisor and other designers

8. Do you know the identity of the planning supervisor?

9. Have you a system for co-operating with the planning supervisor?

10. Do you know the identity of any other designers?

11. Have you a system for co-operating with the other designers?

7   The planning supervisor

Introduction
Definition
Who can be the planning supervisor?
  The client as planning supervisor
  The principal contractor as planning supervisor
Competence
Resources
Requirements on the planning supervisor
The health and safety plan
Basic checklist of considerations for a planning supervisor

Introduction

The Executive has made its views known on many occasions that the role of planning supervisor should not be seen as, nor develop into, a separate profession. However, such a view should not detract from the pivotal role that an effective planning supervisor has on the manner in which the Regulations are implemented. If planning supervisors become better at administering their duties, perhaps the Regulations might deliver some more of the benefits that have been hoped for.

The relatively small number of prosecutions of planning supervisors would suggest that fulfilling the role by 'going through the motions' is sufficient to avoid criticism. However, judging the performance of a planning supervisor is largely subjective except for the clear prescriptive obligations. The clients who appoint the planning supervisor have a genuine difficulty in assessing the quality, as opposed to the competence, of planning supervisors. Identifying how much safer a construction site has been compared with the same site had the planning supervisor not been appointed is truly imponderable.

Definition

Regulation 2(1) provides:

'Planning supervisor' means any person for the time being appointed under regulation 6(1)(a).

The role of planning supervisor is mandatory on all projects which are not exempt from the Regulations. Moreover, the planning supervisor can only be a person appointed by the client by virtue of regulation 6(1)(a) as follows:

Subject to paragraph 6(b), every client shall appoint –

(a)   a planning supervisor;

...in respect of each project.

Once a person has been appointed to the role of planning supervisor it should remain filled until the end of the construction phase, in accordance with regulation 6(5), which provides:

The appointments mentioned in paragraph (1) shall be terminated, changed or renewed as necessary to ensure that those appointments remain filled at all times until the end of the construction phase.

There is no prohibition within the Regulations on terminating, changing or renewing the appointment of persons fulfilling the role of planning supervisor, from time to time. It is suggested that adequate provision is made, in the terms of engagement of the planning supervisor, to enable a smooth handover to the replacement person, particularly with regard to notice of termination or change and identification of documentation to be handed over.

Who can be the planning supervisor?

There is no limitation in the Regulations as to the type of person who may fulfil the role of planning supervisor. The planning supervisor can be an individual, partnership, limited company, local authority or government department subject to demonstrating the necessary competence and allocation of resources to the reasonable satisfaction of the client.

There is no prohibition within the Regulations which prevents the role of planning supervisor being combined with one of the other roles under the Regulations, the main combinations of which are set out below. Combining any role with the planning supervisor will lead to a loss of objectivity and overall experience, from a health and safety management point of view. The practice of appointing an independent planning supervisor under separate terms of engagement and fee arrangements is to be preferred. Unfortunately, the perceived additional cost and extended chain of communication often mitigates against such practice.

The client as planning supervisor

Paragraph (6)(b) of regulation 6 provides:

> *Paragraph (1) does not prevent –*
>
> *(b)   the appointment of the client as planning supervisor...* *provided the client is competent to perform the relevant functions under these Regulations.*

Therefore, a client can appoint itself as the planning supervisor; it does not even have to be 'in-house' as defined within regulation 2(3). The risk of sacrificing the objectivity which an independent planning supervisor would have in fulfilling the requirements under regulation 14 should not be underrated. The lack of objectivity is compounded if the client is also fulfilling one or more of the other roles.

A client has to be reasonably satisfied as to the competence of any person to be appointed as the planning supervisor, as required by regulation 8(1). There is a risk that, despite the express requirement in paragraph (6)(b) that the self-appointing client must be competent to fulfil the role of planning supervisor, the client will not address his mind to the objective assessment of his competence by failing to seek advice from a third party. It is interesting that the client, who appoints himself as the planning supervisor, is not required in paragraph (6)(b) to consider the allocation of resources. This implied waiver of the precondition as to resources before appointing a planning supervisor must be overridden by regulation 9(1) which provides:

> *No client shall appoint any person as planning supervisor in respect of a project unless the client is reasonably satisfied that the person he intends to appoint has allocated or, as appropriate, will allocate adequate resources to enable him to perform the functions of planning supervisor under these Regulations...*

Therefore, a client who appoints himself as planning supervisor should only do so if he is able to allocate, and intends to allocate, adequate resources to the relevant tasks, in addition to being satisfied as to the appropriate level of competence.

The principal contractor as planning supervisor

There is no prohibition in the Regulations which prevents the principal contractor from being appointed as the planning supervisor.

The planning supervisor has, by virtue of regulation 6(3), to be appointed as soon as is practicable after the client has sufficient information to be in a position to ascertain the required competence and allocation of adequate resources by a planning supervisor. However, by virtue of regulation 6(4) a client should only appoint a principal contractor when the client has sufficient information about the project and construction work as will enable him to assess the principal contractor's competence and allocation of resources. If the appointment of the principal contractor is at the same time as the planning supervisor there will not have been the opportunity to prepare a health and safety plan, which would lead to greater difficulty in the client complying with regulations 8(3) and 9(3). The alternative interpretation would be that the client had been too late in appointing a planning supervisor.

If a client wishes to appoint the same person to the roles of planning supervisor and principal contractor at the same time, at the outset of the project and complying with regulations 8(3) and 9(3), it would suggest that the role of principal contractor would not be subject to competitive tendering at a later date. For this reason, it is likely that clients appointing design and build contractors may be more inclined to appoint the design and build contractor to both roles from the outset. For large and complex projects the client should be aware of the risk that the design may result in the principal contractor lacking the competence and resources which might have been thought satisfactory at the beginning. For this reason, a client should keep the issues of competence and resources under review, although the client would be lacking the independent and objective advice of the planning supervisor in such circumstances.

In keeping the appointments of planning supervisor and principal contractor under review, the client can take advantage of combining the roles in the same person at an advanced stage in the pre-construction phase having appointed an independent planning supervisor in the first place to assist with the initial conception and formulation of the project.

A review of the perceived advantages and disadvantages of combining the role of principal contractor and planning supervisor are discussed in Chapter 5 on the appointment of a planning supervisor by the client.

Competence

There has been no profession that has a monopoly on producing planning supervisors from within its ranks. Individuals who are likely to be suited best to the role of planning supervisor will include those who have an understanding of the design process and construction methods. This indicates that civil and structural engineers, architects, builders, and building surveyors are likely to have appropriate backgrounds to enable them to fulfil the requirements of regulation 14. Any person who has attained chartered status in one of the professions previously mentioned will have achieved high academic standards. However, it is unlikely to be sufficient for a planning supervisor to rely on chartered status without evidence of a knowledge of health and safety issues and management. Such evidence can be obtained by individuals taking further qualifications, by examination and assessment in occupational health and safety. However, the Regulations do not make it necessary for a planning supervisor to be of chartered status or have any specialist qualifications. For smaller straightforward projects, the academic and professional qualifications might be satisfactory at the level of technician engineer or engineering technician. Academic and professional qualifications are not, on their own, enough. In addition to the academic qualifications a planning supervisor must be able to demonstrate an appropriate level of experience which may be gained during time spent in an individual's 'first' profession. Ultimately, the required level of academic qualifications for large and complex projects will be of a high standard backed up by several years of varied and relevant experience.

A client seeking to appoint a prospective planning supervisor is required, by virtue of regulation 8(1), to be reasonably satisfied as to the competence of the person. The reasonable enquiries and advice, referred to in regulation 2(5)(a) should be part of the steps taken by a client to ascertain competence. It is suggested that advice should be sought from independent professional sources. These may include professional institutions, learned societies and other persons who offer their services as planning supervisors. A suggested checklist of the enquiries based on the ACOP which a prudent client should make of a person prior to their appointment as a planning supervisor might include:

1.   What academic qualifications does the planning supervisor possess with particular regard to occupational health and safety?

2.   What is the planning supervisor's knowledge of construction practice particularly in relation to the nature of the project? Is it, for instance, building, earthworks, demolition etc.?

3. What is the planning supervisor's familiarity and knowledge of the design function?

4. What is the planning supervisor's knowledge of health and safety issues?

5. Is the planning supervisor able to demonstrate achievement in management, particularly in the role of co-ordination and liaison? Has he the ability to work with and co-ordinate the activities of different designers and be a bridge between the design function and construction work on site?

6. Has the planning supervisor experience of preparing health and safety plans?

7. Does the planning supervisor understand and appreciate the contractual obligations of the other parties to the project?

Resources

The planning supervisor is required to allocate adequate resources to the project, with which the client must be reasonably satisfied, in accordance with regulation 9(1). The enquiries to ascertain the adequacy of resources to be allocated to the project based on the ACOP should include:

1. What number of individuals and their respective qualifications and experience are allocated to the project, both internally and from other sources, to perform the various functions in relation to the project?

2. What management system will be used to monitor the correct allocation of people and other resources in the way agreed at the time these matters were finalised?

3. What management system will be used to co-ordinate the activities of the designers and collect information for the preparation of the health and safety plan?

4. What time will be allowed to key personnel to carry out the different duties of planning supervisor?

5. What technical facilities are available to aid the staff of the planning supervisor in carrying out their duties including:

(a)   Office location and accommodation;

(b)   Information technology;

(c)   Technical library.

6.   Can the planning supervisor demonstrate the implementation and practice of quality assurance principles? Registration by one of the accreditation organisations would be a quick and efficient means of providing the confidence that the designer has a management system incorporating items 2 and 3 above, although such registration is not absolutely necessary.

Requirements on the planning supervisor

The contracts of appointment between clients and planning supervisors extend the duties and obligations beyond those set out in the Regulations. The planning supervisor should be alert to, and avoid if necessary, the additional contractual duties and obligations. What follows is the extent of the statutory duties only.

Once appointed, the planning supervisor records the fact of appointment by notifying the Executive to that effect, by virtue of regulation 7(1) which provides:

> *The planning supervisor shall ensure that notice of the project in respect of which he is appointed is given to the Executive in accordance with paragraphs (2) to (4) unless the planning supervisor has reasonable grounds for believing that the project is not notifiable.*

The planning supervisor is required to consider whether the project is notifiable. If the planning supervisor has reasonable grounds for believing that the project is not notifiable, having addressed his mind to all the relevant matters, he should record in writing those grounds and the information relied upon to make his conclusions. A written record will provide the proof at any later stage that the planning supervisor complied with this regulation. It is suggested, however, that for such grounds to be reasonable, they could not be supported if there was any reasonable doubt that the project was notifiable.

If the planning supervisor considers that the project is notifiable he is required to give the notice, referred to in regulation 7(1) in accordance with regulations 7(2) to 7(4), which provide:

(2) *Any notice required by paragraph (1) shall be given in writing or in such other manner as the Executive may from time to time approve in writing and shall contain the particulars specified in paragraph (3) or, where appropriate, paragraph (4) and shall be given at the times specified in those paragraphs.*

(3) *Notice containing such of the particulars specified in Schedule 1 as are known or can reasonably be ascertained shall be given as soon as is practicable after the appointment of the planning supervisor.*

(4) *Where any particulars specified in Schedule 1 have not been notified under paragraph (3), notice of such particulars shall be given as soon as is practicable after the appointment of the principal contractor and, in any event, before the start of construction work.*

Schedule 1 and a suggested form of notice containing all the information specified in Schedule 1 can be found in Appendix 4.

The planning supervisor is required to serve such a notice on the Executive as soon as possible after his appointment. He should not wait until the time when the principal contractor is appointed. Thus, it may be necessary to serve the notice at least twice or more times on every occasion a contractor is appointed. Note that a notice containing the name and address of the principal contractor must have been served before the start of the construction phase.

The further requirements on the planning supervisor are set out in regulation 14 in paragraphs (a) to (f). Each paragraph sets out a different requirement.

Regulation 14(a) provides:

The planning supervisor appointed for any project shall –

(a) *ensure, so far as is reasonably practicable, that the design of any structure comprised in the project –*

  (i) *includes among the design considerations adequate regard to the needs specified in heads (i) to (iii) of regulation 13(2)(a), and*

  (ii) *includes adequate information as specified in regulation 13(2)(b).*

The planning supervisor is not a designer for the purposes of the Regulations, and should avoid participating in design decisions, while complying with paragraph (a). Since the planning supervisor is not a designer, the role is confined to ensuring a designer has complied with regulations 13(2)(a) and 13(2)(b). Where there is only one designer, the planning supervisor's role is simplified because any dialogue is one to one, but, in situations where there are more than one designer, the planning supervisor's skills of co-ordination and communication become paramount.

Regulation 14(b) provides:

The planning supervisor appointed for any project shall –

(b)   *take such steps as it is reasonable for a person in his position to take to ensure co-operation between designers so far as is necessary to enable each designer to comply with the requirements placed on him by regulation 13.*

In the first instance, the planning supervisor has to identify, for any particular structure, all the designers who may be contributing to the design and other designers whose designs elsewhere on a project may affect or influence the health and safety considerations for the structure. Thus, the planning supervisor becomes the only person on a project in a position to consider how the different aspects of a design and planning will interact. The planning supervisor should ensure the designs of all the designers are considered and reviewed, from a health and safety prospective, to the extent that such designs adopt the general principles of prevention and protection. The principles of prevention to be applied are set out in paragraphs 29–31 of the ACOP to the Framework Regulations and reproduced in Appendix 5.

The means of creating an interactive and integrated flow of information between the different designers can be achieved in a number of ways and will depend on the management style of the planning supervisor. It is only by the information flowing freely between the different parties that the designers will be able to co-ordinate their work and assess how different aspects of the design interact and affect health and safety.

The ACOP draws particular attention to designs carried out by specialist contractors during the construction phase. The planning supervisor has a continuing obligation to ensure that the contractor has complied with regulations 13(2)(a) and 13(2)(b). In these circumstances, the planning supervisor will also need to discuss with the principal contractor and contractor the inclusion of the design into the health and safety plan.

The ACOP suggests that the design of temporary works whilst falling within the scope of the Regulations, will not be a concern for the planning supervisor in all circumstances. Temporary works which are designed to become incorporated in the permanent structure will be subject to the planning supervisor's review. However, general access scaffolds and similar structures which are erected following recognised codes and procedures do not come within the planning supervisor's duties.

Regulation 14(c) provides:

> *The planning supervisor appointed for any project shall –*
>
> *(c)   be in a position to give adequate advice to –*
>
> > *(i)   any client and any contractor with a view to enabling each of them to comply with regulations 8(2) and 9(2); and to*
> >
> > *(ii)   any client with a view to enabling him to comply with regulations 8(3), 9(3) and 10.*

The planning supervisor is not obliged to advise a client or contractor, but should only be in a position to do so if requested. Provided that the planning supervisor is competent and has himself allocated adequate resources to the project he should be able to advise a client or contractor on any aspect of competence. Interestingly, the planning supervisor is not obliged to provide any advice to a designer, perhaps because designers are more likely only to appoint other designers, in which case they will be in the best position to assess competence, although this may not be the situation at all times.

The planning supervisor should be aware that the advice provided to a client or contractor will be a significant factor in satisfying the criteria of 'reasonable satisfaction' before either a client or contractor makes an appointment. The planning supervisor is appointed by the client and, therefore, any advice to a contractor is likely to be outside of any contractual relationship. The planning supervisor should be aware of the common law duty of care to contractors and, for this reason, it is suggested that the advice should be provided in writing, carefully setting out the matters covered by the advice.

The planning supervisor is also required to be in a position to give advice to the client concerning the state of preparation of the health and safety plan. For this reason, if no other, the planning supervisor should establish a flow of information with the principal contractor during the development of the health and safety plan. If a client mistakenly believes that the health and safety plan does comply with regulation 15(4) and intends to approve the commencement of the construction phase the planning supervisor should advise the client accordingly, whether or not his advice has been requested.

Regulation 14(d) provides:

> *The planning supervisor appointed for any project shall –*
>
> (d)   *ensure that a health and safety file is prepared in respect of each structure comprised in the project containing –*
>
> > (i)   *information included with the design by virtue of regulation 13(2)(b), and*
> >
> > (ii)   *any other information relating to the project which it is reasonably foreseeable will be necessary to ensure the health and safety of any person at work who is carrying out or will carry out construction work or cleaning work in or on the structure or of any person who may be affected by the work of such a person at work.*

No other person engaged on a project has the responsibility for ensuring that the health and safety file is prepared and that it fulfils the requirements of paragraph (d). The planning supervisor's responsibility is continuous through to the completion of the construction phase of a project.

The information in the health and safety file will include design information and information added during the construction phase by the principal contractor and other contractors. Information gathered by the planning supervisor and supplied by the client during the preparation of the health and safety plan is also likely to be included in the health and safety file. Refer to Chapter 11 for a detailed discussion of the health and safety file.

Regulation 14(e) provides:

> *The planning supervisor appointed for any project shall –*
>
> (e)   *review, amend or add to the health and safety file prepared by virtue of sub-paragraph (d) of this regulation as necessary to ensure that it contains the information mentioned in that sub-paragraph when it is delivered to the client in accordance with sub-paragraph (f) of this regulation.*

Paragraph (e) is an acknowledgement of the fact that design information, and any other information relating to the project, may change while the principal contractor develops the health and safety plan. It is the planning supervisor's duty to ensure that any changes by the principal contractor or contractors are communicated to him for the purposes of updating the health and safety file. Although the principal contractor is required to supply the planning supervisor with information, by virtue of regulation 16(e), the planning supervisor should put in place review procedures to ensure that the principal contractor is complying with regulation 16(e). Equally, any new information which comes into the possession of the planning supervisor either from his own sources or from the client should be communicated to the principal contractor for the purposes of developing the health and safety plan, and incorporated in the health and safety file.

Regulation 14(f) provides:

> *The planning supervisor appointed for any project shall –*
>
> *(f)   ensure that, on the completion of construction work on each structure comprised in the project, the health and safety file in respect of that structure is delivered to the client.*

Until such time as the construction phase is complete, the planning supervisor has custody of the health and safety file and should take appropriate measures to ensure that it is stored safely with controlled access to it. However, the client takes responsibility for custody of the health and safety file when the construction phase is complete. For discussion on the completion of the construction phase see Chapter 11. It is suggested that the fact of transferring custody of the health and safety file from the planning supervisor to the client is recorded in writing, perhaps by a receipt from the client.

The health and safety plan

Regulation 15(1) provides:

> *The planning supervisor appointed for any project shall ensure that a health and safety plan in respect of the project has been prepared no later than the time specified in paragraph (2) and contains the information specified in paragraph (3).*

The detail and size of the health and safety plan which the planning supervisor has to ensure is prepared, will depend on the nature and extent of the project and the contracting arrangements for the construction work. It is not necessary that the planning supervisor actually prepares the health and safety plan but cannot avoid the responsibility of ensuring that it is prepared. Chapter 10 sets out in detail the requirements of the health and safety plan and the role of the planning supervisor in its preparation.

Basic checklist of considerations for a planning supervisor

1. Have you formal notification from the client of your appointment?

2. Have you notified the Executive of the project?

3. Do you have an agreed line of communication to the client?

Competence and adequate resources

4. Have you demonstrated competence to the client?

5. Have you allocated adequate resources?

6. Are you prepared and able to provide advice to the client and contractors on competence with respect to other members of the project team?

7. Are you prepared and able to provide advice to the client on the allocation of adequate resources by the contractors in the project team?

Health and safety of persons

8. Have you ensured, so far as reasonably practicable, that the design of any structure includes among the design considerations adequate regard to:

   (i)   the avoidance of foreseeable risks?

   (ii) combating risks at source?

   (iii) the giving of priority to measures which will protect all persons at work over measures which only protect each person carrying out work?

9. Have you ensured, so far as reasonably practicable, that the design includes adequate information, to be provided by the designer?

10. Have you taken steps to ensure co-operation between designers?

Health and safety plan

11. Have you ensured that a health and safety plan has been prepared before the invitations to tender have been issued?

12. Have you ensured that the health and safety plan contains the information referred to in regulation 15(3)?

13. Have you handed over the health and safety plan to the principal contractor?

14. Have you a line of communication to the principal contractor for the purposes of contributing to the development of the health and safety plan and the review of the health and safety file?

Health and safety file

15. Have you ensured that a health and safety file is prepared in respect of each structure?

16. Have you kept the health and safety file under review up to completion of the construction phase?

17. Have you handed the health and safety file over to the client on completion of the construction phase?

8   The principal contractor

Introduction

Accidents do not happen in the client's boardroom or in the design office – when the planning and design is done it is the principal contractor who is responsible for the safe execution of the construction works.

The roll call of contracting organisations who have been prosecuted successfully reveals that breach of the Regulations is not confined to the small, badly managed and hardly known contractors. The fact that some of the contractors who have been prosecuted have national and international reputations suggests that close attention to the principal contractor's duties and obligations is still necessary.

Definition

A principal contractor is defined in regulation 2(1) as:

> *...any person for the time being appointed under regulation 6(1)(b).*

The principal contractor is also required to be a contractor by virtue of regulation 6(2), which provides:

> *The client shall not appoint as principal contractor any person who is not a contractor.*

The role of principal contractor is mandatory on all projects which are not exempt from the Regulations. Moreover, the principal contractor can only be appointed by the client by virtue of regulation 6(1)(b) as follows:

> *Subject to paragraph 6(b), every client shall appoint –*
>
> *(b)   a principal contractor,*
>
> *in respect of each project.*

The definition of contractor is discussed in Chapter 9. However, for the purposes of discussing the role of the principal contractor it should be noted that a contractor carries out or manages construction work. Thus a client is able to appoint contractors who do not carry out any construction work but merely provide a management service. This extension of the traditional view of contractors takes account of the trend for management contracts and the use of project management organisations.

Selecting a principal contractor

A client who intends to appoint a principal contractor has to consider two factors, in the same way as a client has before appointing a planning supervisor or designer, notably, competence and resources.

Note that the client is not required to appoint a main contractor or management contractor as the principal contractor. For a discussion on the appointment of other contractors as principal contractor see Chapter 6.

Competence
Regulation 8(3) provides:

> *No person shall arrange for a contractor to carry out or manage construction work unless he is reasonably satisfied that the contractor has the competence to carry out or, as the case may be, manage, that construction work.*

This requirement applies to all contractors, which includes the principal contractor. Therefore, a client has to be 'reasonably satisfied' of a contractor's competence before appointing him as the principal contractor.

The assessment of competence is of variable scope according to the actual role for which the principal contractor is appointed.

First limb

In the first limb of competence, the client needs to be reasonably satisfied as to the contractor's competence to carry out construction work. Not all principal contractors may be appointed to carry out construction work and establishing this first limb of competence to the reasonable satisfaction of the client in such situations does not arise.

Second limb

In the second limb of competence the client needs to be reasonably satisfied as to the contractor's competence to manage construction work. Since it will be the case that a principal contractor will be involved in managing construction work, even if it relates only to the management of health and safety issues on site as between different contractors, a client will always have to consider the contractor's competence with regard to management of construction work. The principal contractor may not be expected to carry out construction work in which case the client's only concern is the contractor's competence to manage construction work.

Both limbs

The principal contractor will often both carry out and manage construction work, in which case the client's enquiries to establish reasonable satisfaction will include both the first and second limbs of competence.

The assessment by the client is not related simply to the ability to carry out or manage construction work, but extends to the principal contractor's awareness, appreciation and ability to create and maintain a safe site. In particular, it is the principal contractor's competence to carry out the duties set out in regulation 16 which is essential.

Adequate resources

Regulation 9(3) provides:

> *No person shall arrange for a contractor to carry out or manage construction work unless he is reasonably satisfied that the contractor has allocated or, as appropriate, will allocate adequate resources to enable the contractor to comply with the requirements and prohibitions imposed on him by or under the relevant statutory provisions.*

To be reasonably satisfied in accordance with regulation 9(3), a client must ascertain, firstly, what resources have been, or are intended to be, allocated by the principal contractor and, secondly, whether such resources are adequate. The planning supervisor is available to advise the client, by virtue of regulation 14(c)(ii).

Selection procedure

During the pre-construction phase, the client will have decided upon the appropriate procurement route for the construction phase and the procedure for selecting the principal contractor. This will either entail a tendering procedure or the client will select one contractor and negotiate the appropriate contract sum and conditions for the construction work.

In the case of selecting a principal contractor by the invitation of tenders, the client should assess the prospective tenderers generally as part of any pre-contract or pre-agreement qualification. If a client undertakes a general assessment of the competence and resources of potential tenderers before the tender stage, it should avoid wasting the time of unsuitable contractors and time spent in assessing such tenders. It is to be hoped that the assessing of potential tenderers at the pre-tender stage will lead to shorter tender lists of more consistent quality, compared with the lengthy tender lists which have become commonplace in recent times.

When the tenders have been received, or one contractor has been selected for a negotiated contract, it should be possible to assess the contractor's competence and resources against the more specific requirements relating to the project. The health and safety plan will be a key document in making the appropriate checks for competence and resources.

The enquiries that may be made by a client, whether at the pre or post tender stage, as part of the reasonable steps taken to satisfy the requirements of 'reasonable satisfaction', and based on the ACOP, include the following:

1. What is the contractor's health and safety policy including details of the policy review procedure?

2. What arrangements does the contractor have in place to manage health and safety, both at head office and on site?

3. What procedures will the contractor adopt for developing and implementing the health and safety plan?

4. What approach will the contractor take to deal with the high risk areas identified by designers and the planning supervisor?

5.   What arrangements does the contractor have for monitoring compliance with health and safety legislation?

6.   How many persons will be allocated to carry out and/or manage the construction work? What are the skills and qualifications of such persons?

7.   What time has the contractor allowed in the construction programme to complete the various stages of the construction work without risks to health and safety?

8.   In what way will persons be employed to ensure compliance with site safety procedures, including details of site induction training programmes?

9.   What means will the contract adopt by which occupational health and safety advice is provided to personnel?

10. What methods does the contractor use on existing sites to involve employees in the health and safety procedures and review process?

11. What are the reported accident frequency rates for all UK construction activities in which the contractor is involved? This information should be provided for a certain period of time together with an analysis of such accidents.

12. What convictions, if any, has the contractor or any of its employees, in respect of health and safety legislation over a certain period of time?

In addition to any enquiries as to competence or resources the client may make of the principal contractor, he may seek the advice of the planning supervisor who is available to provide such assistance by virtue of regulation 14(c)(ii).

Requirements on and powers of the principal contractor

The requirements to which the principal contractor is subject and his powers are set out in regulation 16. Regulation 16(1), which also happens to be one of the most frequent grounds for prosecution, lists such requirements and powers as follows:

> *The principal contractor appointed for any project shall –*
>
> *(a)* *take reasonable steps to ensure co-operation between all contractors (whether they are sharing the construction site for the purposes of regulation 11 of the Management of Health and Safety at Work Regulations 1999 or otherwise) so far as is necessary to enable each of those contractors to comply with the requirements and prohibitions imposed on him by or under the relevant statutory provisions relating to the construction work;*

The principal contractor's obligation to take reasonable steps to ensure co-operation between all contractors will include his own domestic sub-contractors, nominated sub-contractors and other contractors employed directly on the project by the client, or any other person, e.g. utility companies. The role under regulation 16(1)(a) is to manage and give practical effect to the duties upon all contractors engaged on the construction site to ensure an integrated approach to health and safety management on site.

Regulation 11 of the Framework Regulations addresses specifically the duties of employers and self-employed sharing a workplace to co-ordinate their activities, co-operate with each other and share information to help each comply with the duties under the relevant statutory provisions. In particular, the relevant statutory provisions include the Framework Regulations and the Work Equipment Regulations.

The steps that are reasonable for the principal contractor to take depend on the scale and complexity of the project. At the least, they will involve a procedure for communicating information, notably the programming and construction intentions of each contractor.

Every contractor is required, by virtue of regulation 3 of the Framework Regulations, to prepare a risk assessment which should address risks to employees and to other persons who may be affected by their activities, such as members of the public. Therefore, the principal contractor will prepare a risk assessment as will every other contractor. Three aspects of the risk assessments prepared by the other contractors on a project will influence the role of the principal contractor. These are:

1. The seriousness of the risk.

2. The nature of the assessment. The risk assessments prepared by other contractors are likely to comprise two parts. According to the normal activities of the contractor there is likely to be a generic risk assessment which will not require any amendment from project to

project. However, for each individual project the second part of the
risk assessment will need to be tailored to the particular risks for each
project. In certain cases, the project may have certain risks of such a
nature that the contractor would be required to prepare a fresh risk
assessment.

3.   The principal contractor will be concerned to examine the
      inter-relationship between the risk assessments produced by the
      other contractors, including his own. Some activities of a contractor
      will have no effect on other contractors working on the project,
      whereas in certain circumstances activities of a contractor will affect
      one or more other contractors to a greater or lesser extent.

The health and safety plan, which identifies the risks, should be
reflected in the risk assessments of the other contractors. The principal
contractor should confirm that the seriousness of the risks has been evalu-
ated properly by the other contractors. Most importantly, it is the principal
contractor's responsibility to identify where problems arise from each contractor's approach to eliminating or reducing such risks.

The ACOP suggests that in circumstances where a number of contractors may be exposed to the same risk, the principal contractor could
co-ordinate the preparation of a single risk assessment. Such a risk assessment would be undertaken for and on behalf of the other contractors and
they should approve the approach to the risk assessment so as to be satisfied
that regulation 3 of the Framework Regulations has been fulfilled.

Health and safety arrangements

The principal contractor may identify an interaction problem arising from
either the risk assessments produced by the other contractors or from a
common assessment undertaken by himself on behalf of the other contractors. In this situation it is the principal contractor's responsibility to
ensure that the general principles of prevention and protection are applied
to measures to be taken, which will eliminate or reduce the risks. The
principles of preventive and protective measures are set out in paragraphs
29–31 of the ACOP to the Framework Regulations and reproduced in
Appendix 5.

The principal contractor would also be required, in accordance with
regulation 5 of the Framework Regulations to 'give effect to such arrangements as are appropriate'. Therefore, the agreed measures and arrangements arising out of the risk assessments must deal with the risk, or risks,
in a co-ordinated and effective way. The principal contractor should either

undertake the appropriate action arising out of the risk assessment or allocate such action to one or more of the other contractors. Since the analysis of the risk assessments should take place before the construction phase the agreed measures and arrangements should be incorporated into the health and safety plan as part of its development.

Health surveillance
The principal contractor should ascertain the provisions for health surveillance by every contractor to determine whether it is a matter which should be co-ordinated. For instance, individual contractors may be able to fulfill the requirements of health surveillance, in accordance with regulation 6 of the Framework Regulations whereas the employees of other contractors may be exposed to risks for which their employers lack the resources for health surveillance. The principal contractor may choose to introduce a system common to all contractors on the project or set up arrangements between selected contractors. The arrangements adopted by the principal contractor should be included in the health and safety plan, so that all contractors on the project are aware of the arrangements which might apply to them and to others.

Health and safety assistance
Every contractor is required, in accordance with regulation 7 of the Framework Regulations to, 'appoint one or more competent persons to assist him in undertaking the measures he needs to take to comply with the requirements and prohibitions imposed upon him by or under the relevant statutory provisions'. Some of the contractors on a project may have the resources to appoint competent persons, but the likelihood of all contractors having such resources on large and complex projects is doubtful. In any event, where there is interaction between the contractors, even if each has the resources to appoint the necessary competent persons, the principal contractor may decide to co-ordinate the provision of health and safety assistance. Whatever approach the principal contractor adopts to co-ordinate health and safety assistance, the arrangements should be included in the health and safety plan.

Procedures for serious and imminent danger and for danger areas
Emergency procedures, dealt with by regulation 8 of the Framework Regulations, will in most projects require a co-ordinated approach. The principal contractor is likely to undertake the role of co-ordinating emergency procedures except where the construction work is being carried out on a client's operating premises. A client having construction work done on its own operating premises is likely to have its own emergency procedures

which should be adopted or modified to take account of the various contractors and activities. In circumstances where the principal contractor takes a leading role in preparing the emergency procedures he is responsible for ensuring that relevant information is exchanged. Note that the emergency procedures should be communicated to all employees working on the construction site and all other persons visiting the site under the control of the principal contractor or any other contractor.

Common equipment
The principal contractor and other contractors have duties under regulation 4 of the Work Equipment Regulations to ensure that work equipment is safe and that it is used safely. Thus, where work equipment is shared by a number of contractors, for instance generators, scaffolding, cranage etc., its provision and use should be co-ordinated by the principal contractor in accordance with regulation 11 of the Framework Regulations.

The responsibility for ensuring that work equipment is safe and that it can be used safely is likely to depend upon the contractor who is supplying the equipment. In the case of cranes, the contractor providing the crane is required to comply with all the relevant statutory provisions. However, the principal contractor may decide to exercise a co-ordination role for the benefit of other contractors on the site by checking and verifying that such statutory provisions have been observed and communicating the fact to the other contractors. Conversely, the provision of cranage on a particular project may be of central importance and the principal contractor would take responsibility directly for such equipment which may be used by other contractors.

The ACOP stresses the importance of the need for co-operation and exchanging information between contractors, to ensure that faults or changes in the conditions of use are reported to the co-ordinator (either the principal contractor or a contractor designated for such purposes) for the equipment. By this means, instructions or limitations on use should be passed to the common users.

The health and safety plan should include all details regarding control, co-ordination and management of shared equipment so that no contractor should be in any doubt as to who has responsibility.

Contractors cannot divest themselves of their own legal responsibilities simply on the basis that the principal contractor may have assumed a responsibility to co-ordinate certain aspects of health and safety management or has undertaken to implement common arrangements. Each contractor remains responsible for satisfying himself that he is not in breach of any of the statutory provisions. In certain circumstances, particularly with regard to specialist contractors, the peculiar risks of that contractor's activities are such that the principal contractor might insist upon that contractor providing his own arrangements despite the fact that there might be duplication.

In the event of any failure in the common arrangements which resulted in a breach of the statutory provisions, the apportionment of liability between the principal contractor, any contractor directed by him to perform a particular task and any other contractors would be decided on the facts of each case.

Regulation 16(1)(b) provides:

The principal contractor appointed for any project shall –

> *(b)* *ensure, so far as is reasonably practicable, that every con-tractor, and every employee at work in connection with the project complies with any rules contained in the health and safety plan;*

As the health and safety plan is developed by the principal contractor he will establish rules for all persons and contractors on the construction site, for the purposes of implementing the measures and arrangements arising out of the various risk assessments.

As the first step in ensuring that every contractor on the project com-plies with any rules contained in the health and safety plan, the prin-cipal contractor should ensure that every contractor has a copy of the health and safety plan complying with regulation 15(4). Any amendment to the health and safety plan during the construction phase should be notified to all contractors by the principal contractor. In respect of bringing to the attention of every employee at work in connection with the project the rules contained in the health and safety plan, the principal contractor should be satisfied with, and monitor, the other contractors' communication of information and training in accordance with regulation 17. Equally, the principal contractor must take steps to ensure that his own employees comply with any rules contained in the health and safety plan.

The duty on the principal contractor is only to ensure 'so far as is reason-ably practicable' that every person complies with the rules. Such steps which are reasonably practicable will depend upon the degree of risk. Thus, in appropriate circumstances it would be reasonable to test regularly the employees exposed to certain risks and provide full time surveillance and supervision to ensure compliance.

Regulation 16(1)(c) provides:

The principal contractor appointed for any project shall –

(c)   take reasonable steps to ensure that only authorised persons are allowed into any premises or part of premises where construction work is being carried out;

An authorised person is not defined by the Regulations, although the ACOP provides some guidance. Authorised persons are authorised by the principal contractor or the client, either individually or collectively. The authorisation will either allow them to enter all, or only specified parts, of the area where construction work is taking place. Authorised persons may include contractors or named employees carrying out construction work and any other persons who need access to such areas including designers and other visitors. Note that persons who have a statutory right to enter an area where construction work is being carried out are deemed to be authorised persons.

The principal contractor will need a procedure for identifying authorised persons as a prerequisite to exclude non-authorised persons. What steps are reasonable for the principal contractor to take to exclude non-authorised persons from work areas will depend on all the circumstances, including the nature of the project and the location.

The ACOP suggests that measures to control access must be related to what is foreseeable. The comparison between the measures needed for a large remote site compared with an urban site close to a school illustrates how measures to control access vary and in the latter case need to be more secure and comprehensive.

In addition to the principal contractor's duty under the Regulations to control access to work areas, the contractor in possession of the site, which may or may not be the principal contractor, owes a duty to lawful visitors by virtue of the Occupier's Liability Act 1957 who may not have been authorised. Additionally, the Occupier's Liability Act 1984 governs the duty of a contractor in possession of a site as to the safety of persons who are outside the scope of the 1957 Act. Persons covered by the 1984 Act include those who are on a construction site without the contractor's permission, whether with lawful authority or without. Thus, special consideration will be needed for protecting:

(i)   persons exercising rights of way across sites and sites which are part of, or adjacent to, other work areas; and

(ii)   trespassers without lawful authority to be on the site.

Regulation 16(1)(d) provides:

The principal contractor appointed for any project shall –

(d)   *ensure that the particulars required to be in any notice given under regulation 7 are displayed in a readable condition in a position where they can be read by any person at work on construction work in connection with the project;*

By virtue of regulation 7 the planning supervisor is required to send a notice to the Executive containing the information set out in Schedule 1 to the Regulations, which can be found in Appendix 4.

A copy of the notice sent by the planning supervisor, or other notice containing the same information, should be displayed so that it can be read by those working on and affected by the site. The principal contractor should select a position, or positions, on the site which will come to the attention of all persons at work and such that it remains in a readable condition. The principal contractor should ensure that the notice remains legible at all times. In addition to the displaying of the notice, the principal contractor should ensure that all contractors are aware of the contents of the notice to enable them to comply with their duties under regulations 19(2) to (4).

Regulation 16(1)(e) provides:

The principal contractor appointed for any project shall –

(e)   *promptly provide the planning supervisor with any information which –*

    (i)   *is in the possession of the principal contractor or which he could ascertain by making reasonable enquiries of a contractor, and*

    (ii)   *it is reasonable to believe the planning supervisor would include in the health and safety file in order to comply with the requirements imposed on him in respect thereof in regulation 14, and*

    (iii)   *is not in the possession of the planning supervisor.*

The requirement on the principal contractor, by virtue of this regulation, provides the necessary communication link to the planning supervisor for the purpose of ensuring that the health and safety file is complete. The principal contractor should be alert to any design or other changes initiated by himself or other contractors which might affect the information referred to in regulation 13(2)(b) or other information referred to in regulation 14(d)(ii). Such information in the possession of the principal contractor has to be provided promptly to the planning supervisor. However, the principal contractor cannot rely on the other contractors informing him of information which may need to be communicated to the planning supervisor. The principal contractor may be unaware of any additional information or change to existing information as a result of the decisions or actions of the other contractors. In such circumstances, how is the principal contractor to know when to make reasonable enquiries? The principal contractor would be advised to instigate a review of the state of information by the making of regular enquiries from every contractor.

Regulation 16(2) provides:

> *The principal contractor may –*
>
> *(a)* *give reasonable directions to any contractor so far as is necessary to enable the principal contractor to comply with his duties under these Regulations;*
>
> *(b)* *include in the health and safety plan rules for the management of the construction work which are reasonably required for the purposes of health and safety.*

The power of the principal contractor to give directions to other contractors is solely to enable him to comply with his duties under the Regulations. The principal contractor has no power to give directions to other contractors for the purposes of enabling him to comply with duties under other regulations or statutory provisions.

The principal contractor's powers to give directions to his own sub-contractors can be provided for in the sub-contract conditions of contract. Therefore, the sub-contractors are under a commercial pressure to comply with the principal contractor's reasonable directions. However, similar commercial pressures may be difficult to achieve in respect of other contractors who have no direct contractual link with the principal contractor. Ideally, the client and the planning supervisor should ensure that all contractors are subject to a contractual obligation to comply with the reasonable directions of the principal contractor. If the reasonable directions can be anticipated by the principal contractor before the commencement of the construction phase they can be incorporated as rules in the health and safety plan. This would avoid the potential for claims from other contractors for additional expense arising from compliance with reasonable directions, of which there had been no prior notice. Note, however, that other contractors are required to comply with the reasonable directions of the principal contractor, by virtue of regulation 19(1)(c), but only on the basis that such directions are reasonable.

The health and safety plan should include details of the means by which reasonable directions of the principal contractor will be communicated to other contractors and how compliance will be achieved and monitored.

Regulation 16(3) provides:

> *Any rules contained in the health and safety plan shall be in writing and shall be brought to the attention of persons who may be affected by them.*

The principal contractor's duty under this regulation is self-explanatory. However, it should be noted that the requirement to bring such rules to the attention of persons who may be affected by them is absolute. There is no 'relaxation' to take all reasonable steps.

Information and training

Regulation 17(1) provides:

> *The principal contractor appointed for any project shall ensure, so far as is reasonably practicable, that every contractor is provided with comprehensible information on the risks to the health or safety of that contractor or of any employees or other persons under the control of that contractor arising out of or in connection with the construction work.*

The principal contractor should review the health and safety plan to ensure that it contains information which will be comprehensible to the contractors and provide them with the relevant parts. Thus, if the health and safety plan is stored on a computer disk, rather than in writing, the principal contractor should produce a copy in writing to those contractors that would be unable to access the information. Special consideration would be required where contractors and their employees are unable to speak English or it is not their first language. The principal contractor should ensure, so far as is reasonably practicable, that there is a contractual obligation on every contractor to hand over a copy of the health and safety plan to other contractors they may be appointing as sub-contractors.

Ideally, the health and safety plan should be made available to contractors when they tender for the different work packages of the construction phase. By this means, contractors will be in a better position to determine the allocation of adequate resources, in order to comply with the requirements of regulation 9(3).

Regulation 17(2) provides:

> *The principal contractor shall ensure, so far as is reasonably practicable, that every contractor who is an employer provides any of his employees at work carrying out the construction work with –*
>
> (a) *any information which the employer is required to provide to those employees in respect of that work by virtue of regulation 10 of the Management of Health and Safety at Work Regulations 1999; and*
>
> (b) *any health and safety training which the employer is required to provide to those employees in respect of that work by virtue of regulation 13(2)(b) of the Management of Health and Safety at Work Regulations 1999.*

The information which an employer is required to provide to its employees by virtue of regulation 10 of the Framework Regulations includes information on risks and on preventative and protective measures (limited to what employees need to know to ensure their own health and safety). The regulation also requires information to be provided on the emergency arrangements, including the identity of staff nominated to assist in the event of evacuation. The information required by virtue of regulation 10 should be included in the health and safety plan and should, therefore, be the basis for the provision of information to be provided by the principal contractor.

Every employer is required to provide his employees with adequate health and safety training at the start of their employment and on their being exposed to new or increased risks thereafter. Similarly, the arrangements for provision of training by employers should be set out in the health and safety plan, although the details of the training will be a matter for each contractor with regard to the employee's existing capabilities, level of training, knowledge and experience.

If a principal contractor sought to comply with regulation 17(1) and 17(2) without providing copies of the health and safety plan, complying with regulation 15(4), to all other contractors, in the light of the guidance provided by the ACOP it might be very difficult for him to demonstrate that he had taken the steps, so far as is reasonably practicable, to ensure compliance.

Advice from, and views of, persons at work

Regulation 18 provides:

> *The principal contractor shall –*
>
> (a)　*ensure that employees and self-employed persons at work on the construction work are able to discuss, and offer advice to him on, matters connected with the project which it can reasonably be foreseen will affect their health or safety; and*
>
> (b)　*ensure that there are arrangements for the co-ordination of the views of employees at work on construction work, or of their representatives, where necessary for reasons of health and safety having regard to the nature of the construction work and the size of the premises where the construction work is carried out.*

Where a recognised trade union has appointed representatives under the Safety Representatives and Safety Committees Regulations 1977, the principal contractor should work with them and their committees. Employers have a duty to consult safety representatives under regulation 4A of the Safety Representatives and Safety Committees Regulations and to provide reasonable facilities and assistance.

On large sites, the principal contractor may adopt a two-tier approach to arrangements for consultation over health and safety issues. At the first tier, the principal contractor might organise an overall safety committee for the site comprising the safety representatives for each contractor. The second tier would be the safety committees for each contractor, with a representative from the principal contractor in attendance, as and when required in the opinion of the principal contractor. The essential requirement in any management system for consultation, is the provision of the opportunity for the principal contractor to consult with employees and the self-employed, to listen and take action, if appropriate, on advice provided by them.

On sites where there are no recognised trade unions, or safety representatives have not been appointed, or there is incomplete coverage of all persons, the principal contractor will need to make other arrangements for consultation. The arrangements will be tailored to the size and nature of the project, so as to enable co-ordination of the views of employees working for different contractors to be taken into account.

Basic checklist of considerations for a principal contractor

1. Have you formal notification from the client of your appointment?

2. Has the planning supervisor notified the Executive of your appointment?

3. Do you have an agreed line of communication to the client?

4. Do you have an agreed line of communication to the planning supervisor?

Competence and adequate resources

5. Are you competent?

6. Are you reasonably satisfied as to the competence of any contractors or designers engaged by you?

7. Have you sought the advice of the planning supervisor with regard to competence before making the appointments?

8. Have you allocated adequate resources?

9. Are you reasonably satisfied as to the allocation of adequate resources by any contractors or designers engaged by you?

Health and safety plan and file

10. Have you received a health and safety plan?

11. Have you completed the development of the health and safety plan including the introduction of any rules?

12. Have you provided a copy of the health and safety plan to the client and received confirmation that the construction phase can start?

13. Have you made arrangements for the management and monitoring of compliance by all persons with the health and safety plan?

14. Have you taken steps to bring to the attention of all persons the rules within the health and safety plan?

15. Have you supplied a copy of the health and safety plan to every contractor or supplied them with comprehensible information on the risks to the health or safety of that contractor?

16. Have you made arrangements to communicate to the planning supervisor any information he might need for the health and safety file?

17. Have you obtained the planning supervisor's agreement to changes in the health and safety plan due to design changes?

Notification of health and safety measures

18. Have you ensured that every contractor who is an employer has provided his employees with the information required on emergency procedures, risks and preventative and protective measures for their own health and safety?

19. Have you ensured that every contractor who is an employer has provided his employees with health and safety training?

20. Have you ensured that employees and self-employed persons are able to bring to your attention advice or any other matters connected with the project which can reasonably be foreseen will affect their health and safety?

21. Have you ensured that there are arrangements for co-ordinating the views of all employees at work on matters of health and safety?

Contractors

22. Are you aware of the identity of all contractors on the project?

23. Have you taken reasonable steps to ensure co-operation between all contractors in accordance with regulation 16(1)(a)?

The site

24. Have you taken reasonable steps to ensure that only authorised persons are allowed on to the construction site?

25. Have you ensured that the notice given under regulation 7 is displayed on the site in a prominent position?

9 Contractors

Definition
Selecting a contractor
Requirements and prohibitions on contractors
Basic checklist of considerations for contractors

Definition

A contractor is defined in regulation 2(1) as:

> *...any person who carries on a trade, business or other undertaking (whether for profit or not) in connection with which he –*
>
> *(a)   undertakes to or does carry out or manage construction work,*
>
> *(b)   arranges for any person at work under his control (including, where he is an employer, any employee of his) to carry out or manage construction work.*

A contractor is concerned with construction work, the definition of which has been reviewed in Chapter 3.

In sub-paragraph (a) a contractor is a person who 'undertakes to or does'. Thus, although a person may not actually be carrying out or managing construction work, such a person is deemed to be a contractor if he is under some obligation to do so.

Both sub-paragraphs (a) and (b) refer to carrying out or managing construction work. A contractor may be engaged, therefore, in only one of the activities.

Sub-paragraph (b) relies on the concept of 'arranges'. The class of person subject to control under the contractor's arrangement is those who are 'at work'. This is an important distinction because a person who is under the control of another person but is not 'at work' does not bring that person within the definition of 'contractor'. The phrase 'at work' is not defined. The narrower definition of 'at work', meaning actual execution of 'carrying out or managing construction work', is to be preferred to a wider definition which could include persons who may be subject to the control of a person while carrying out work, which is different from the work of 'carrying out or managing construction work'.

In the simple case of an employer arranging for employees of his to carry out or manage construction work, the employer is deemed to be a contractor. However, there are other employers once removed from the execution of carrying out or managing construction work who can be deemed to be contractors under the Regulations. The concept of 'arranges' is set out in regulation 2(2), which provides:

> *In determining whether any person arranges for a person (in this paragraph called 'the relevant person') to prepare a design or carry out or manage construction work regard shall be had to the following...*
>
> (a)   *a person does arrange for the relevant person to do a thing where –*
>
>   (i)   *he specifies in or in connection with any arrangements with a third person that the relevant person shall do that thing (whether by nominating the relevant person as a sub-contractor to the third person or otherwise)*

Note in this case that the relevant person will 'carry out or manage construction work'. Provided that the person who arranges for the relevant person to do the thing in accordance with (i), and retains control over the relevant person whilst at work, the person is deemed to be a contractor. Therefore, although a client may arrange for a nominated sub-contractor to be sub-contracted to a contractor (the third person) the lack of control over the nominated sub-contractor whilst at work keeps the client outside the definition of contractor. Conversely, a client with his own direct labour force who specifies in a contract with a contractor that the direct labour force will carry out certain construction work will be deemed to be a contractor under the Regulations by virtue of the control which the client would have over its direct labour force at work.

In the context of management contracting, the client will have direct contracts with the works contractors who are, in turn, required by an arrangement with the management contractor to do certain work. In this situation the client has a direct contractual control over the works contractors while at work. Therefore, the client, in these situations, is deemed to be a contractor unless he appoints the management contractor as his agent.

A further concept of 'arranges' is provided by paragraph (2)(a)(ii) which provides:

> *(a)*　　*a person does arrange for the relevant person to do a thing where –*
>
> *(ii)*　　*being an employer, it is done by any of his employees in-house*

Note that, by virtue of paragraph (3)(b) of regulation 2, employees have only to be engaged in 'a separate part of the undertaking of the employer' to be treated as 'in-house'. It is not essential that the employees are in a subsidiary company or separate partnership. Thus, a large manufacturer which has a maintenance unit would be deemed to be a contractor where the maintenance unit became involved in construction work that falls within the scope of the Regulations.

Paragraph (2)(b) of regulation 2 identifies certain situations where a person does not arrange for the relevant person to do a thing.

The Regulations clarify further the scope of the term 'arranges' by defining situations which do not come within the definition of arranges in paragraph (2)(b) of regulation 2 as follows:

> *a person does not arrange for the relevant person to do a thing where –*
>
> *(i)*　　*being a self-employed person, he does it himself or, being in partnership it is done by any of his partners; or*

The circumstances in (i) are easy to identify. A self-employed person is the only possible person to be deemed a contractor. In the case of partnerships, the partners are jointly and severally liable for all the responsibilities and liabilities of the partnership. Therefore, a partner asking another partner to do a thing is the equivalent of doing the same thing himself.

> *(ii)*　　*being an employer, it is done by any of his employees otherwise than in-house; or*

That is to say that the employees do not form 'a separate part of the under-taking of the employer' distinct from the part for which the design is prepared. Companies do not 'arrange', as defined, because the employees of such companies will not, in most cases, be a separate part of the company's undertaking. Thus, a contractor that has an internal management organisation in which the temporary works designers work alongside the estimators would not be arranging within the definition of the Regulations.

>   (iii) *being a firm carrying on its business anywhere in Great*
>   *Britain whose principal place of business is in Scotland, it is*
>   *done by any partner in the firm; or*

This situation is self-explanatory.

>   (iv) *having arranged for a third person to do the thing, he does not*
>   *object to the third person arranging for it to be done by the*
>   *relevant person;*
>
>   *and the expressions 'arrange' and 'arranges' shall be construed*
>   *accordingly.*

Thus, a contractor who appoints a design and build sub-contractor is not
deemed to have arranged for the preparation of a design if the design and
build sub-contractor arranges, in turn, for another person (i.e. the relevant
person) to prepare the design.

Selecting a contractor

By virtue of regulations 8(3) and 9(3), set out in full in Chapter 4 on
Competence and resources, any person arranging for a contractor to carry
out or manage construction work has to be reasonably satisfied as to that
contractor's competence and allocation of adequate resources.

Matters which a person should consider before appointing a contractor
in accordance with regulation 2(5) are essentially the same as those matters
contained in the suggested checklist for the principal contractor in
Chapter 8.

Requirements and prohibitions on contractors

Regulation 19(1)(a) provides:

>   *Every contractor shall, in relation to the project –*
>
>   (a) *co-operate with the principal contractor so far as is necessary*
>   *to enable each of them to comply with his duties under the rel-*
>   *evant statutory provisions*

This regulation is the vital all-embracing obligation on a contractor to
co-operate. Without such co-operation it would become impossible for the

principal contractor to comply with his requirements under regulations 16(1)(a), (b) and (e).

Regulation 19(1)(b) provides:

> *Every contractor shall, in relation to the project –*
>
> *(b)   so far as is reasonably practicable, promptly provide the principal contractor with any information (including any relevant part of any risk assessment in his possession or control made by virtue of the Management of Health and Safety at Work Regulations 1999) which might affect the health or safety of any person at work carrying out the construction work or of any person who may be affected by the work of such a person at work or which might justify a review of the health and safety plan*

A contractor is under a duty to notify the principal contractor, as soon as possible, of any information which becomes known to him which might affect the health and safety of at least two categories of persons. The first category of persons are those carrying out construction work and secondly, the general public including those persons who are on adjacent sites or who are exercising a lawful right of way through or alongside the construction site. Whilst there is no express requirement for a contractor to notify the principal contractor of any information which affects the health or safety of persons engaged in maintenance or cleaning work after the construction phase is concluded, these are matters which might justify a review of the health and safety plan. In any event, a contractor should have a copy of the health and safety plan and is under a continuing duty to inform the principal contractor of any matter which comes to the knowledge of the contractor and may affect the contents of the health and safety plan.

Regulation 19(1)(c) provides:

> *Every contractor shall, in relation to the project –*
>
> *(c)   comply with any directions of the principal contractor given to him under regulation 16(2)(a)*

Provided that the directions of the principal contractor are limited to enabling him to comply with his duties under the Regulations, a contractor is bound, by virtue of this regulation, to comply with such directions. This will be so regardless of any contractual relationship or otherwise which might exist between the principal contractor and contractor.

Regulation 19(1)(d) provides:

Every contractor shall, in relation to the project –

(d) comply with any rules applicable to him in the health and safety plan

It is a matter for the principal contractor's discretion whether or not to include in the health and safety plan rules for the management of the construction work. In circumstances where the principal contractor has made such rules and they are written in the health and safety plan, the contractor is bound to comply with such rules in so far as they are applicable to him. Note that the principal contractor has a duty to bring such rules in the health and safety plan to the attention of a contractor who may be affected by them by virtue of regulation 16(3).

Regulation 19(1)(e) provides:

Every contractor shall, in relation to the project –

(e) promptly provide the principal contractor with the information in relation to any death, injury, condition or dangerous occurrence which the contractor is required to notify or report by virtue of the Reporting of Injuries, Diseases and Dangerous Occurrences Regulations 1995

The information which the contractor is required to compile and notify or report to the enforcing authorities, by virtue of the Reporting of Injuries, Diseases and Dangerous Occurrences Regulations 1995, should be promptly provided by delivering copies of such information to the principal contractor. The principal contractor needs the information to monitor compliance with health and safety law and ensure that the arrangements for the management of health and safety remain appropriate throughout the duration of the construction phase.

Regulation 19(1)(f) provides:

Every contractor shall, in relation to the project –

(f) promptly provide the principal contractor with any information which –

(i) is in the possession of the contractor or which he could ascertain by making reasonable enquiries of persons under his control, and

> *(ii)   it is reasonable to believe the principal contractor would provide to the planning supervisor in order to comply with the requirements imposed on the principal contractor in respect thereof by regulation 16(1)(e), and*
>
> *(iii) which is not in the possession of the principal contractor.*

This regulation is remarkably similar to the obligation on the principal contractor, by virtue of regulation 16(1)(e), to provide information to the planning supervisor. In order to comply with 'making reasonable enquiries of persons under his control' the contractor would be well advised to instigate a regular review system of construction work under his control, possibly through the means of a health and safety committee.

Regulations 19(2) and (3) provide:

> *(2)   No employer shall cause or permit any employee of his to work on construction work unless the employer has been provided with the information mentioned in paragraph (4).*
>
> *(3)   No self-employed person shall work on construction work unless he has been provided with the information mentioned in paragraph (4).*

These regulations complete the 'cascade' of information down to every person who is, or is likely, to work on construction work. The information referred to in paragraphs (2) and (3) and set out in paragraph (4) is as follows:

> *(a)   the name of the planning supervisor for the project;*
>
> *(b)   the name of the principal contractor for the project; and*
>
> *(c)   the contents of the health and safety plan or such part of it as is relevant to the construction work which any such employee or, as the case may be, which the self-employed person, is to carry out.*

Regulation 16(1)(d) requires the display of the notice given under regulation 7, which contains the names of the planning supervisor and the principal contractor. However, the duty on a contractor is higher than the duty merely to display the notice because a contractor is required to provide the information to employees and self-employed persons.

Regulation 17(1) requires that every contractor should be provided with comprehensible information. The relevant information will normally be contained in part of the health and safety plan. Thus it is the responsibility of a contractor to identify those parts which affect particular employees and self-employed persons for the purposes of taking copies and providing them to the employees and self-employed persons.

Regulation 19(5) provides:

> *It shall be a defence in any proceedings for contravention of paragraph (2) or (3) for the employer or self-employed person to show that he made all reasonable enquiries and reasonably believed –*
>
> *(a)   that he had been provided with the information mentioned in paragraph (4); or*
>
> *(b)   that, by virtue of any provision in regulation 3, this regulation did not apply to the construction work.*

To avoid the accusation that the information referred to in paragraph 4 has not been provided to employees or self-employed persons, the information should be provided in writing at the commencement of the construction phase or the person's employment, receipt of which is acknowledged by the signature of the employee or self-employed person and retained in the contractor's records.

Note that for this regulation not to apply the client must have reasonable grounds for believing that the project is not notifiable and that the largest number of persons at work, at any one time, will not exceed four. In all other cases a planning supervisor will have been appointed and therefore a notice, to be given under regulation 7, will be in existence. The employer or self-employed person cannot simply plead ignorance in any defence to breach of paragraphs (2) or (3). They are under a duty to make reasonable enquiries of the principal contractor or planning supervisor as to the existence of such information. The information which is provided to the employer or self-employed person should be checked by them to identify any obvious omissions or errors otherwise they would not be in a position to rely on reasonable belief that the information had been provided in accordance with paragraph (4).

Basic checklist of considerations for contractors

1. Do you have an agreed line of communication to the principal contractor?

Competence and adequate resources

2. Are you competent?

3. Are you reasonably satisfied as to the competence of any other contractors engaged by you?

4. Have you sought the advice of the planning supervisor with regard to competence before making such appointments?

5. Have you allocated adequate resources?

6. Are you reasonably satisfied that any contractors engaged by you have allocated adequate resources?

Health and safety plan and file

7. Have you received the health and safety plan?

8. If you have not received the health and safety plan have you made reasonable enquiries to ascertain the relevant information?

9. Are you prepared and able to comply with any rules applicable to you in the health and safety plan?

10. Have you provided the principal contractor with a risk assessment or any other information which may affect the health and safety plan?

11. Have you provided to the principal contractor any notice by virtue of the Reporting of Injuries, Diseases and Dangerous Occurrences Regulations 1995?

12. Have you provided to the principal contractor any information which is not in the possession of the principal contractor and which he is bound to provide to the planning supervisor?

13. Have you delayed the start of any work until receipt of the relevant information?

Professional team

14. Have you received the names of all members of the professional team, i.e. planning supervisor, principal contractor, client, designers, etc.?

15. If you have not received the names of the professional team have you made reasonable enquiries to ascertain such information?

16. Are you prepared and able to co-operate with the principal contractor?

17. Are you prepared and able to comply with any directions of the principal contractor?

10 The health and safety plan

Definition
Background
Stage one: Preparation of the plan
  *Who is responsible for the initial preparation of the health and
    safety plan?*
  When should the health and safety plan be ready?
  What should be included in the health and safety plan?
Stage two: Development of the plan
  Development of the health and safety plan by the principal contractor
Health and safety plan at completion of the construction phase

Definition

The health and safety plan is defined in regulation 2 as:

> *...the plan prepared by virtue of regulation 15.*

Regulation 15 sets out the requirements of the health and safety plan and
therefore, providing the requirements are observed and incorporated
within a plan, it is, for the purposes of the Regulations, the health and
safety plan. Note that the health and safety plan is in respect of any project,
where the Regulations are deemed to apply, whereas the health and safety
file, discussed in Chapter 11, is in respect of each structure within a project.

The health and safety plan does not have to be a written document,
except for any rules, by virtue of regulation 16(3). It could be stored on
a computer database, which would offer particular advantages to contributors and the handover between the planning supervisor and the principal contractor. In most cases, of course, the health and safety plan will
be written document and it is recommended, in the case of computer
storage, that a hard copy of the plan is made at key stages of development
for ease of reference by those who may not have the necessary viewing
facilities.

Background

The health and safety plan, at one and the same time, is the touchstone and continuous management thread for any project. Its preparation commences from the moment that a planning supervisor is appointed, and is completed when it complies with all the requirements of regulation 15, although it remains subject to continuous review and amendment in the light of circumstances which may have changed after the start of construction.

The health and safety plan is a unique document in so far as there is no other document, in the project documentation commonly used, in which every party to a project has a stake in its final form – and effectiveness. Essentially, the health and safety plan is a medium for communication. Figure 5 illustrates the basic arrangement for contributing to the health and safety plan during its preparation and development.

The success in improving the management of health and safety in construction relies heavily on the effectiveness of the health and safety plan. It compels each party to a project to consider health and safety in a way which no other document could do. As the health and safety plan is developed the rationale adopted by the various contributors should be obvious. The 'transparency' of the health and safety plan and its effectiveness is also one of the measures that the Executive are able to use as part of their policing function. It is often the first source of information that any health and safety inspector would wish to see on visiting a site or investigating an accident. The health and safety plan is likely to be vital evidence in the Executive's decision making process when considering enforcement action or prosecution. Equally, in pursuance of any action in civil proceedings for injury or death, the health and safety plan will be important evidence.

Stage 1: Preparation of the health and safety plan

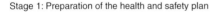

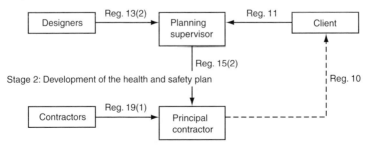

Figure 5 Contributors to the health and safety plan. The dahed line indicates referral back to the client before the start of the construction phase

Stage one: Preparation of the plan

Who is responsible for the initial preparation of the health and safety plan?

Regulation 15(1) provides:

> *The planning supervisor appointed for any project shall ensure that a health and safety plan in respect of the project has been prepared no later than the time specified in paragraph (2) and containing the information specified in paragraph (3).*

The planning supervisor is in fact only responsible to ensure that the plan has been prepared, which begs the question as to whether this amounts to a responsibility to prepare the plan before it is provided to the principal contractor. Some uncertainty is introduced by the guidance in the ACOP which refers to the planning supervisor having to prepare and develop the health and safety plan.

The client has a duty to provide information to the planning supervisor, by virtue of regulation 11, which information will be included either in part, or in its entirety, in the health and safety plan. The client relies upon the planning supervisor to utilise the information and has no involvement in the health and safety plan except for the purposes of ensuring that it complies with regulation 15(4) before construction work starts (regulation 10).

Designers are required to co-operate, in accordance with regulation 13(c), with the planning supervisor and other designers to 'enable each of them to comply with the requirements and prohibitions' imposed by the Regulations. This includes the requirement on the planning supervisor to ensure that the health and safety plan has been prepared. Thus, whilst the Regulations do not require expressly that the planning supervisor shall prepare the health and safety plan, this appears to admit the possibility that the preparation of the health and safety plan could be delegated by the planning supervisor or client to another party. Ultimately, whether or not the planning supervisor actually prepares the health and safety plan, it remains his responsibility to ensure the plan has been prepared, however it is done.

When should the health and safety plan be ready?

Regulation 15(2) provides:

> *The time when the health and safety plan is required by paragraph (1) to be prepared is such time as will enable the health and safety*

> *plan to be provided to any contractor before arrangements are*
> *made for the contractor to carry out or manage construction work.*

The planning supervisor should ensure that the health and safety plan is available to any contractor, as soon as possible before the award of a contract. The ACOP suggests that the health and safety plan should be part of the tender documentation. The planning supervisor is not prevented from handing over sections of the health and safety plan, despite the fact it may not be completed. Details of some structures may be completed which will enable the contractor to develop such parts of the health and safety plan before others. Handing over the health and safety plan to the contractor in sections may be a means of saving time, thus helping to achieve an earlier start on site.

What should be included in the health and safety plan?

The detail and size of the health and safety plan will depend on the nature and extent of the project and the contracting arrangements for the construction work. Clearly, the appearance of the health and safety plan included in the tender documentation for a design and build contract will be very different from the health and safety plan which might be prepared for tenderers in the 'traditional' procurement route.

Regulation 15(3) sets out the information which is required to be in the health and safety plan in sub-paragraphs (a) to (f) inclusive, before it is handed over to the principal contractor.

Regulation 15(3)(a) provides:

> *a general description of the construction work comprised in the*
> *project;*

In its simplest form the health and safety plan would include a general description of the works which would correspond with the description of the works in the main contract documentation. However, this may be insufficient if key elements of a project are being tendered separately. This would be so because the contract description of the works in each case will not describe the whole project.

Since the health and safety file has to be compiled in respect of each structure within the project it would be advisable to describe the construction works in the health and safety plan by reference to each structure to ensure the project is adequately described.

Regulation 15(3)(b) provides:

> *details of the time within which it is intended that the project, and*
> *any intermediate stages, will be completed;*

The planning supervisor's enquiries to ascertain the intended completion date, or period within which construction must be completed, should involve the client in an honest and frank disclosure to the planning supervisor concerning the client's real needs. However, the client's requirements may be tempered by information from the designers which will include assessments of construction duration. At the end of the planning supervisor's enquiries, the client should be appraised of the factors which will influence the completion date for the project, not least of which will be cost. The planning supervisor's enquiries should have identified issues arising out of a consideration of health and safety matters, which the client will need to consider, in finalising a programme for the project including earlier completion of certain structures. The earlier dates for completion of certain structures or sections of the project, if appropriate, correspond to the 'intermediate stages'.

Note also that the completion of the design phase may be an intermediate stage if the construction phase commences before the design phase is completed in its entirety.

Regulation 15(3)(c) provides:

> *details of risks to the health or safety of any person carrying out the construction work so far as such risks are known to the planning supervisor or are reasonably foreseeable;*

Note that the risks are those known to the planning supervisor. Just in case the planning supervisor does not possess sufficient knowledge, the requirement is qualified as including details of risks which are 'reasonably foreseeable': foreseeable by whom? The interpretation most likely to be adopted would be the reasonable foreseeability attributable to an ordinarily competent planning supervisor thus reverting to an objective assessment as compared with the subjective assessment if it had been related solely to the planning supervisor.

Regulation 15(3)(d) provides:

> *any other information which the planning supervisor knows or could ascertain by making reasonable enquiries and which it would be necessary for any contractor to have if he wished to show –*
>
> (i) *that he has the competence on which any person is required to be satisfied by regulation 8, or*
>
> (ii) *that he has allocated or, as appropriate, will allocate, adequate resources on which any person is required to be satisfied by regulation 9;*

   The information falls into two categories, the first of which is any other information in the knowledge of the planning supervisor or information obtained by making reasonable enquiries for the purpose of enabling any contractor to demonstrate that it has the necessary competence, in accordance with regulation 8, to satisfy the person making the appointment.

   The second category of information is similar to the first, except that it is such information as will enable any contractor to demonstrate it has allocated, or will allocate, adequate resources, in accordance with regulation 9, to satisfy the person making the appointment.

   The adequacy of the information required may be judged according to the use to which it can be put by a contractor to demonstrate competence and the necessary allocation of resources. Thus, if the information provided by the planning supervisor is solely within his own knowledge and is insufficient to enable a contractor to demonstrate competence and adequate resources, the planning supervisor is on notice of the need to make reasonable enquiries to supplement that knowledge. A difficulty may arise in the situation where an inexperienced planning supervisor is unaware of what information is appropriate and is unable to judge whether the information in his own knowledge is sufficient. There is no reason why, even after making reasonable enquiries, the information provided to a contractor should be less than sufficient. If there is a lack of sufficient information due to the 'incompetence' of the planning supervisor, that would be a matter of concern to the client with regard to his duty to appoint a competent planning supervisor, by virtue of regulation 6(a).

   Regulation 15(3)(e) provides:

> *such information as the planning supervisor knows or could ascertain by making reasonable enquiries and which it is reasonable for the planning supervisor to expect the principal contractor to need in order for him to comply with the requirement imposed on him by paragraph (4);*

The planning supervisor has to know, subject to the test of reasonableness, what a principal contractor will need in order to comply with regulation 15(4). In other words, the planning supervisor has to provide sufficient information to the principal contractor to enable him to ensure that the health and safety plan has the features specified in regulation 15(4). The planning supervisor is not required to double guess every possible approach by the principal contractor to the work involved in the construction phase, but is required to make reasonable assumptions as to what would or would not be needed by the principal contractor.

Note again that the planning supervisor is obliged to make reasonable enquiries if he has reason to believe or suspect that information within his own knowledge would be inadequate for the principal contractor's purposes. Regulation 15(3)(f) provides:

> *such information as the planning supervisor knows or could ascertain by making reasonable enquiries and which it would be reasonable for any contractor to know in order to understand how he can comply with any requirements placed upon him in respect of welfare by or under the relevant statutory provisions.*

This last requirement is particularly interesting because it is the only one which relates to the information which a contractor needs, to comply with the relevant statutory provisions concerning the welfare of its employees and others for whom it is responsible. The 'relevant statutory provisions' will include the C(HSW) Regulations with regard to toilet and washing facilities, drinking water, accommodation for clothing, facilities for changing clothes and facilities to rest and eat meals. At the least, the health and safety plan should identify specific welfare facilities that may be available, including shared facilities.

On certain sites a client may have its own safety rules to protect the health and safety of workers, on chemical plants and nuclear installations in particular. Such rules should be included under this heading even if there has not been a reason to include them under any previous heads of requirements.

The health and safety plan should not merely be a list of particular risks and a source of information which a contractor will need, but should indicate when appropriate or necessary, what precautionary measures should be adopted. A client may wish, for its own reasons, to have a contractor adopt a specific approach to the implementation of measures in response to a particular risk. The designer's assessment of risk, which may not be obvious to a contractor, may dictate specific precautions which the designer would ensure are communicated to the planning supervisor. Finally, other legislation or guidance issued by authoritative organisations may require or recommend respectively the adoption of particular precautionary measures.

Every health and safety plan will be unique and it is entirely within the judgment of the planning supervisor as to what should or should not be included. A suggested checklist of matters which every planning supervisor should ensure has been addressed, regardless of whether the result identifies a risk, is set out below. It would advisable for a planning supervisor to identify the source of all the information contained within the health and safety plan.

General

1. Name and address of the client.

2. Name and address of the planning supervisor.

3. Name and address of the designer(s).

4. Statement of the client's health and safety policy.

5. Description of the works and identification of specific structures.

Programme

6. Project completion date or period of time during which the site for the works is available.

7. Identification of phases or sections and respective interim completion dates.

Existing off-site conditions

8. Land use adjacent to the site with particular regard to dangerous buildings, industrial processes and the release of noxious substances.

9. The potential for contaminated groundwater flow from adjacent land.

10. The potential for ground instability arising from movement on adjacent land.

11. Transitory hazards including the transportation of liquid petroleum gas, radioactive materials and ammunition.

12. Traffic systems and restrictions including waiting and delivery times.

Existing on-site considerations

13. Existing services including gas, electricity and water.

14. Traffic systems on, under or over the site.

15. Traffic restrictions affecting waiting and delivery times.

16. Site investigation data and ground conditions identifying potential instability from old and current mineworkings, underground obstructions, past industrial/chemical contamination, release of methane and other gases and risk of spontaneous combustion.

17. Existing structures (as defined in the Regulations) to be described together with the particular risks which might arise from demolition or refurbishment. This information should include, but is not limited to:

    (a) Identifying noxious substances or materials, e.g. asbestos, fuels etc.

    (b) Pre and post tensioned members.

    (c) Load bearing limits and other reasons for potential instability.

Existing records

18. Available drawings of structures to be demolished, refurbished or incorporated in the new works.

19. Health and safety file, if available, for any structures on the site.

20. Site investigation reports.

21. Historical maps and records with particular regard to mining, flooding and weather.

The design

22. Principles and assumptions of design for structures comprised in the construction work identifying, in particular, states of instability. Further information might include the identification of any precautions that are needed or any method or sequence of assembly that should be followed during construction.

23. Detailed reference to specific risks where contractors will be required to explain and present their proposals for minimising or avoiding such risks.

24. Hazards or work sequences identified as being of particular risk to the health and safety of construction workers and which cannot be eliminated.

Construction materials

25. Health hazards from construction materials specified in the contract documentation (having been selected because there is no suitable alternative). If possible, the construction materials should be identified with particular structures or phases of the work.

Site layout and management

26. Positions of site access and egress.

27. Location of storage areas and unloading arrangements.

28. Location of temporary site accommodation.

29. Layout or positions of welfare facilities.

30. On-site traffic and pedestrian routes including zoning where different considerations for health and safety might apply.

Relationship with the client's undertaking

31. Identification and consideration of the health and safety issues which arise when the project is located in premises occupied or partially occupied by the client.

Site rules

32. Security requirements, if any.

33. Specific site rules with which the client or planning supervisor requires compliance.

34. Specific site rules which are required by other statutory provisions.

Procedures for the continuing review of the health and safety plan

35. Procedures for considering the health and safety implications of elements designed by the principal contractor and/or other contractors.

36. Procedures for dealing with unforeseen eventualities and conditions during the construction phase resulting in substantial design change which might affect resources.

Stage two: Development of the plan

At an appropriate time the health and safety plan, prepared under the auspices of the planning supervisor, is handed over to the principal contractor to complete its development, as provided for by regulation 15(4):

> *The principal contractor shall take such measures as it is reasonable for a person in his position to take to ensure that the health and*

safety plan contains until the end of the construction phase the following features

Regulation 5(2) requires the health and safety plan to be provided to any contractor 'before arrangements are made for the contractor to carry out or manage construction work'. The ACOP states that the health and safety plan should be prepared sufficiently for it to form part of the tender documentation. A tendering contractor needs all the information referred to in regulation 15(3) to be able to price the work that will be necessary to comply with the health and safety requirements. To be as meaningful as possible, the planning supervisor should be satisfied that the health and safety plan has been developed to a stage which includes all the information referred to in regulation 15(3) immediately prior to tendering, subject to handing over the health and safety plan in sections.

The time required by a principal contractor to develop a health and safety plan will depend on the size and complexity of the project. It will also depend on the ease with which the principal contractor can find and assimilate the information contained within the plan. Presentation of the health and safety plan is an important skill. Each section of the health and safety plan should be clearly identified by subject matter or structures, and the use of lists and diagrams should be encouraged. Sources of information should be identified to enable the principal contractor to save time by approaching such sources directly. Despite the fact that the principal contractor will at some time after the appointment have responsibility for developing the health and safety plan it is unlikely that the planning supervisor will cease to have any further contribution. It is envisaged in the ACOP that there will be discussion and liaison between the planning supervisor and principal contractor to ensure that the transition is as smooth as possible.

The extent of co-operation needed between the planning supervisor and the principal contractor in design and build projects will be considerable. The same organisation may be responsible for design and construction or it may be different organisations, each responsible for design or construction respectively. In either set of circumstances, the planning supervisor should develop the health and safety plan as far as possible before the tender documents or similar are prepared. As the detailed design of the project proceeds, the planning supervisor will need to prepare the health and safety plan but in close co-operation with the principal contractor. Thus, the design and build principal contractor will not be given responsibility for developing the plan until the planning supervisor is satisfied that the design has been advanced to a stage where the information referred to in regulation 15(3) has been obtained.

Development of the health and safety plan by the principal contractor

The 'features', referred to in regulations 15(4)(a) and (b), which a principal contractor is required to develop and include within the health and safety plan are as follows:

(a) *arrangements for the project (including, where necessary, for management of construction work and monitoring of compliance with the relevant statutory provisions) which will ensure, so far as is reasonably practicable, the health and safety of all persons at work carrying out the construction work and all persons who may be affected by the work of such persons at work, taking account of –*

 (i) *risks involved in the construction work,*

 (ii) *any activity specified in paragraph (5); and*

(b) *sufficient information about arrangements for the welfare of persons at work by virtue of the project to enable any contractor to understand how he can comply with any requirements placed upon him in respect of welfare by or under the relevant statutory provisions.*

If the planning supervisor has included the features in the health and safety plan before responsibility passes to the principal contractor, the principal contractor is not under an obligation to include them! Nonetheless, even if such features are included, the principal contractor may not be satisfied that they are sufficiently developed or free from error. It would be prudent to assume that a principal contractor has a duty to check such features which may have been included prior to taking responsibility for the health and safety plan, and therefore check their adequacy or suitability against his own plans for the construction work.

The arrangements which the principal contractor is required to set out in the health and safety plan are for the purpose of demonstrating, so far as it is reasonably practicable, that all persons involved in the construction process and 'all persons who may be affected by the work' are not exposed to any undue risks to their health and safety. Therefore, the scope of persons which the principal contractor has to consider will include visitors to the site, persons on adjacent sites and properties and the public at large. Note that persons, other than workers on the project, may be a possible hazard to the health and safety of the workers. An example would be a chemical plant next to a construction site which may release chemicals across the construction site.

The principal contractor's first consideration of the arrangements must take account of the risks involved in the construction work, which will be identified in the health and safety plan in accordance with regulation 15(3). Secondly, the principal contractor must take account of the activities which will be involved in the project. The definition of an activity is set out in regulation 15(5) which provides:

> (5) An activity is an activity referred to in paragraph (4)(a)(ii) if –
>
> > (a) it is an activity of persons at work; and
> >
> > (b) it is carried out in or on the premises where construction work is or will be carried out; and
> >
> > (c) either –
> >
> > > (i) the activity may affect the health or safety of persons at work carrying out the construction work or persons who may be affected by the work of such persons at work, or
> > >
> > > (ii) the health or safety of the persons at work carrying out the activity may be affected by the work of persons at work carrying out the construction work.

This definition is self-explanatory although it should be noted that it includes activities 'carried out in or on the premises' before construction work commences. This would include site inspections and survey work prior to the start of construction which could involve entry by persons into sewers and inspection of unstable cliffs as two typical examples.

In planning the construction work the principal contractor should identify all the activities. This will be assisted by the preparation of method statements for each job explaining how the contractor intends to construct the works, each of which is likely to comprise numerous activities.

Paragraph (4)(a) requires that the arrangements should include, where necessary, 'management of the construction work and monitoring of compliance with statutory provisions'. Therefore, the arrangements may be developed in the primary sense on a site wide basis and in the secondary sense as applied to activities. A principal contractor who sets out the features required in accordance with paragraph (4)(a) should include the following, which can be considered as a basic checklist:

1. The approach to be adopted for managing health and safety by every-one involved in the construction phase. This will include, in addition to the principal contractor and his employees, all other contractors, designers, suppliers delivering to site and any other category of person involved in or contributing to the work in the construction phase.

2. Assessments produced by contractors in accordance with regulation 3 of the Framework Regulations and other relevant legislation. The activities dealt with in the C(HSW) Regulations provide a useful checklist, see Appendix 6.

3. Arrangements for monitoring compliance with health and safety law, including inspection regimes, qualifications of staff and recording.

4. Rules, where appropriate, for the management of the work for health and safety including notices, posters or other means of communicating such rules.

5. Common arrangements, which may be imposed by the client or imposed by the principal contractor, e.g. wheel washing, traffic management etc.

6. Arrangements for the performance and fulfilment of the principal contractor's duties under regulations 16–18.

The principal contractor is also required to include the information referred to in paragraph (4)(b) concerning the welfare of persons at work. The following should also be included by the principal contractor in the health and safety plan as part of the checklist.

7. Common welfare arrangements which may be provided by the client or the principal contractor.

Arrangements for monitoring the health and safety plan as the construction phase proceeds should be included by the principal contractor. The need to make modifications may arise from:

1. The receipt of information from other contractors who are obliged to provide such information in accordance with regulation 19.

2. The need to make design changes.

3. The occurrence of unforeseeable circumstances.

4.   The principal contractor wishing to change some or all of the princi-
     ples on which the health and safety plan was prepared by the planning
     supervisor.

5.   Revision to existing, or the introduction of new health and safety
     legislation.

6.   Revision to existing, or the introduction of new industry standards.

Note that the principal contractor should consult with and obtain the
approval of the planning supervisor to any variations in the health and
safety plan which have any design implications arising from unforeseen
eventualities before implementing such variations.

Health and safety plan at completion of the construction phase

At the completion of the construction phase, the purpose of the health and
safety plan will hopefully have been fulfilled, in so far as it stimulated a
greater awareness of the hazards and risks to health and safety and a
concomitant improvement in communication between all the parties to the
project. However, the Regulations do not provide for the safekeeping of
the health and safety plan although it would be advisable for a principal
contractor to retain the plan together with, and for as long as, all the other
contract documentation is retained.

The planning supervisor has a duty to compile a health and safety file
under regulation 14 in respect of each structure, and may request a copy of
the health and safety plan from which certain parts will be extracted for
inclusion in the health and safety file.

11 The health and safety file

Definition

There is no regulation which is concerned solely with the health and safety file. The health and safety file is defined in regulation 2(1) as:

> *...a file, or other record in permanent form, containing the information required by virtue of regulation 14(d).*

The definition makes it clear that the health and safety file does not have to be in any particular format provided that it is in a 'permanent form'. Thus, photocopies or faxes which are prone to fade should not be included in any health and safety file.

If the health and safety file is kept on a magnetic record, it would not be permanent if it is susceptible to inadvertent erasure unless the contents of the file are stored on a separate disc under adequate security measures against such inadvertence.

The only guidance provided by ACOP is that the health and safety file should be easy to store securely and facilitate easy retrieval of information.

Who is responsible for preparing the health and safety file?

Regulation 14(d) provides:

> *The planning supervisor appointed for any project shall –*

143

> *(d)   ensure that a health and safety file is prepared in respect of each structure comprised in the project...*

It is not essential that the planning supervisor prepares the health and safety file. The planning supervisor may delegate the preparation of the file to other persons better able or suited to the task.

The planning supervisor's responsibility for compiling the health and safety file is limited to ensuring that the information complies with the requirements in paragraphs (i) and (ii) of regulation 14(d) and that it is in a permanent form.

The planning supervisor also has further obligations set out in regulation 14(e) which provides:

> *review, amend or add to the health and safety file prepared by virtue of sub-paragraph (d) of this regulation as necessary to ensure that it contains information mentioned in that sub-paragraph when it is delivered to the client in accordance with sub-paragraph (f) of this regulation.*

Thus, the planning supervisor should ensure that he is made aware of any changes to any aspects of the project, including individual structures, during the construction phase which may alter the health and safety plan and the basis of information for the health and safety file.

Project versus structure

Since a project may have more than one structure, as defined in the Regulations, it follows from regulation 14(d) that there will be more than one health and safety file for many projects. Thus, quite straightforward building projects are likely to have at least two health and safety files; typically, in respect of the building itself and the temporary works 'to provide support or means of access during construction'. It is not difficult to envisage projects where numerous structures can be identified. For example, a highway project might comprise the following structures:

- Bridges;
- Sewers;
- Earth retaining structures;

in addition to the road itself.

The planning supervisor would have a responsibility to ensure that a health and safety file for each of the above structures is prepared. However, the ACOP suggests it may be appropriate and practical for there to be one health and safety file prepared for a project comprising a number of structures. In such circumstances the health and safety file for each structure would consist of the contents of the health and safety file for the project which are relevant to that structure.

Effectively, the health and safety file for each structure would be a chapter, section or sub-file of the health and safety file for the project, which does not detract from the responsibility to prepare a 'file' for each structure.

What information should be included in the health and safety file?

Regulation 14(d) refers to the information which should be included in the health and safety file, the first part of which provides:

> (i)   *information included with the design by virtue of regulation 13(2)(b)...*

Reference to regulation 13(2)(b) reveals that the information should include:

> *...adequate information about any aspect of the project or structure or materials (including articles or substances) which might affect the health and safety of any person at work carrying out construction work or cleaning work in or on the structure at any time or of any person who may be affected by the work of such a person at work.*

All of this information should be included with the design, including the design of means of access for maintenance and further construction work once the project has been completed. Note that the health and safety file will identify those aspects of cleaning and maintenance work which might affect the health and safety of the persons carrying out such work and those who may in consequence be affected. Tenants and others who have an interest in projects after construction will be particularly interested with regard to information on cleaning work.

The planning supervisor should be careful to ensure that all designs prepared by all the designers contributing to the project are collected together for the purpose of gathering all the information in existence by virtue of regulation 13(2)(b).

The second part of regulation 14(d) provides:

> (ii) *any other information relating to the project which it is reasonably foreseeable will be necessary to ensure the health and safety of any person at work who is carrying out or will carry out construction work or cleaning work in or on the structure or of any person who may be affected by the work of such a person at work.*

Thus, all other information, not included with the design, which might affect the health and safety of the persons identified in regulation 14(d)(ii) has to be included in the health and safety file. This further information will be provided primarily by:

(i) The principal contractor and other contractors pursuant to regulation 16(1)(e);

(ii) The client pursuant to regulation 11; and

(iii) Information from within the planning supervisor's own knowledge and resources.

Note that the information is not related to persons who will be using the facilities of the project except to the extent that they may be at risk as a result of construction or cleaning work.

A non-exhaustive checklist of the items of information which should be included in the health and safety file, based upon the ACOP, is set out below.

1. Historic site data.

2. Site survey information, pre and post construction phase.

3. Site investigation reports and records.

4. Photographic record of essential site elements.

5. Statement of design philosophy, calculations and applicable design standards.

6. Drawings and plans used throughout the construction process, including drawings prepared for tender purposes.

7. Record drawings and plans of the completed structure.

8.   Maintenance instructions.

9.   Instructions on the handling and/or operation of equipment together with the relevant maintenance manuals.

10. The results of proofing or load tests.

11. The commissioning test results.

12. Materials used in the structure identifying, in particular, hazardous materials including data sheets prepared and supplied by suppliers.

13. Identification and specification of in-built safety features, for example, emergency and firefighting systems and fail safe devices (see also item 9 above).

14. Method statements produced by the principal contractor and/or contractors.

What happens to the health and safety file on completion of the construction work?

On completion of the construction work, the planning supervisor's obligations are set out in regulation 14(f) which provides:

The planning supervisor appointed for any project shall –

(f)   ensure that, on the completion of construction work on each structure comprised in the project, the health and safety file in respect of that structure is delivered to the client.

The reference to 'completion of construction work' is not defined, nor is any guidance provided in the ACOP. Most standard forms of contract for building or civil engineering works contain expressions for completion such as 'substantial completion' under the ICE Conditions and 'practical completion' under JCT 98. These expressions are also not defined with any certainty, except to the extent that the issue of a certificate by the appropriate person under the relevant contract is evidence of such completion. However, in most projects, there follows a period of time during which the contractor has a continuing obligation to complete minor items, maintain the project and remedy any defects. There is a significant likelihood that a contractor will continue to be involved in construction work and it is suggested therefore that the planning supervisor should not hand over the health and safety file to the client until the expiry of the maintenance or defects liability period.

On receipt of the file by the client, it should be kept safe and treated as though it were part of the title documents for the project and available for use at any time in the future as envisaged by the Regulations. The client's obligations to maintain safe custody of the file are set out in regulation 12, see Chapter 5.

12 Project documentation

Introduction
Clients and agents
The professional appointments – duty of care
Terms of engagement of the planning supervisor
Terms of engagement of designers
Contractors' tender documentation
Standard conditions of contract
  JCT Standard Form of Building Contract 1998 Edition
  ICE Conditions of Contract (Measurement Version) 7th Edition
  MF/1 (Revision 4) Model Form of General Conditions of Contract

Introduction

There is no need to include in a contract, as between the parties to a project, the precise obligations and duties which are set out fully in the Regulations. The requirements and prohibitions as they affect the roles of contracting parties according to the Regulations apply regardless of any contractual arrangement. However, for the sake of clarity and avoiding confusion there are various matters that should be considered when preparing documentation for projects to which the Regulations apply.

An unplanned benefit of the Regulations associated with compelling the different parties to communicate with each other, has been to import an improved understanding and clarity of the other party's intentions. The Regulations have probably done as much for an improvement in contractual relationships on projects as they have done for health and safety.

Some of the relevant matters for consideration to be included in the project documentation are dealt with below according to the various roles under the Regulations.

Clients and agents

It will not be uncommon for any number of persons to be associated with the inception of a project, any of whom could be identified as a client.

These persons may include lenders, landowners, developers, joint venture members, future purchasers or tenants. Accordingly, the documentation which will record the agreement between the various persons will range from funding agreements to agreements for a lease. To avoid any doubt as to which person will fulfil the role of client for the purposes of the Regulations the relevant person should be identified in the relevant documentation. The person selected as client should be required to make a declaration in accordance with regulation 4(4). The other persons should consider whether they want an indemnity from the 'client' should any claim arise from the client's breach of the Regulations. Conversely, the other persons may wish to have a cause of action against the selected client for any claim flowing from his breach of the Regulations. Note that an indemnity cannot provide protection against the risk of a criminal prosecution.

Clients appointing agents are advised to make such appointments in writing recording precisely the extent of authority given to the agent. The agent should be required to make a declaration in accordance with regulation 4(4) and indemnify the client against the consequences which would arise from failing to make such a declaration, or breaching any of the requirements and prohibitions in the Regulations.

The agent should be given full powers of authority to fulfil the role in accordance with the Regulations. If the agent is in any way fettered in performing the role to the full, he is exposed to a risk of prosecution over which he may have no control.

The client should give some consideration to the consequences of an event, such as insolvency, which would interrupt or disable the agent from fulfilling his role.

In addition, the client should provide for terminating the appointment of an agent in the event of a specified breach of agency. Changing the agent or client during the course of a project is not prohibited although it should be remembered that unless there is a fresh declaration in accordance with regulation 4(4) the last person notified to the Executive remains the client under the Regulations.

The professional appointments – duty of care

In the case of the planning supervisor and designers they will be obliged to perform their respective duties in accordance with the Regulations to avoid the risk of a criminal prosecution. Therefore, in their relationship with the Executive the obligation to perform as required by the Regulations is absolute. The Executive does not have to demonstrate any loss and there is no possibility of a defence which relies on demonstrating they had performed at least as well as any other person reasonably skilled in their particular field of expertise. However, planning supervisors and designers are not likely to be in a position to give such an absolute warranty of compliance with the Regulations to clients and their appointers respectively.

The professional indemnity insurers will not, in most if not all cases, provide cover for obligations owed to clients by professionals which go beyond the duty of care established in the case of *Bolam -v- Friern Hospital Management Committee.*

Thus, most clients will have to accept that the contractual obligation owed by planning supervisors or designers will be limited to a duty of care expected of a professional in the same field of expertise as the one in which they are practising. Therefore, the consequences for a planning supervisor or designer arise from two concurrent liabilities, one of which is criminal and the other a contractual duty of care.

A breach of the Regulations may result in a criminal conviction without any consequences arising from a breach of contract if there has been no loss to the client. Conversely, it is possible that a failure to perform the requirements and observe the prohibitions under the Regulations may not lead to a criminal conviction (in the discretion of the Executive), although the client may have suffered a loss for which damages are recoverable in civil proceedings.

At the present time, standard forms of contract, which provide for the contractor to design and build, limit the contractor's liability for design to the duty of care expected of a professional in the relevant discipline. Whilst this duty of care for design is unlikely to be disturbed by the Regulations it will be a matter for negotiation whether a duty of care should be applied to the role of planning supervisor, in circumstances where this role is undertaken by the principal contractor or contractor.

It is to be anticipated that all collateral warranties given in the future by professionals, and contractors, will reflect the new obligations under the Regulations, in the same terms as the main appointments. Even the planning supervisor may be asked to give collateral warranties to provide comfort for lenders, purchasers or tenants that the health and safety file has been prepared in accordance with the Regulations.

Terms of engagement of the planning supervisor

The client has to appoint a planning supervisor by virtue of regulation 6(1)(a). An executed agreement, incorporating the terms of engagement for the appointment of a planning supervisor would be evidence of the appointment.

The terms of engagement should state expressly that the person is appointed as planning supervisor in accordance with the Regulations. The duties of the planning supervisor set out in the Regulations cannot be varied by contract and are deemed to be incorporated.

In negotiating the terms of engagement of the planning supervisor, some of the matters which might be considered include the following:

1. Warranties as to competence and the allocation of adequate resources.

2. The timing and manner of the provision of information by the client.

3. Confirmation as to who has the responsibility for preparing the health and safety plan.

4. Requests for advice by the client.

5. The timing and manner of the handing over of the health and safety file.

6. Provisions for terminating the appointment of the planning supervisor with particular regard to handing over.

7. Reporting regime by the planning supervisor of information to the client.

8. Fee arrangements (time charge or percentage basis).

Planning supervisors should be alert to the tendency in various standard and bespoke forms of engagement to impose wider obligations than are required by the Regulations.

Terms of engagement of designers

The designer is bound at all times by the relevant duties and obligations contained within the Regulations. Accordingly, there is no requirement to set out in any terms of engagement the requirements of regulation 13 which the designer is obliged to undertake in any event.

There are certain matters which, it is advised, should be included in the terms of engagement of a designer. These include:

1. The identity of the planning supervisor and a provision for a subsequent appointment.

2.   The identities of any other designers known at the time.

3.   The identity of the principal contractor if known at the time, and a provision for a subsequent appointment.

4.   Confirmation as to who has the responsibility for preparing the health and safety plan.

5.   A notice, by virtue of regulation 13(1), to the client of his obligations under the Regulations.

6.   Details of any notices pursuant to regulation 7 or any declarations pursuant to regulation 4, if appropriate.

7.   Warranty by the designer as to competence and the allocation of adequate resources.

Contractors' tender documentation

All tender documentation sent to tendering contractors and sub-contractors for projects subject to the Regulations, in addition to the conditions of contract, bills of quantities and specification or employer's requirements (as appropriate), should include the following information:

1.   Confirmation, or otherwise, that the Regulations apply.

2.   The notice in accordance with regulation 7 including:

      (i)   The identity of the planning supervisor.

      (ii) The identity of the designer(s).

      (iii) The identity of the principal contractor (if appointed), or confirmation that the award of the contract will be the appointment of the principal contractor in accordance with regulation 6(1)(b).

3   The declaration in accordance with regulation 4(4), if appropriate.

4.   The health and safety plan.

   The ACOP envisages that the health and safety plan prepared during the pre-construction phase should be provided to the tendering contractors with the tender documentation, see regulation 15(2). Unless the health and safety plan has been developed to a stage that would allow a tendering contractor to price fully the tender, the client will be exposed to the risk of a late revision to tender prices or later contractual claims. Even if the health and safety plan is not finalised at the start of the tendering process, it would be advisable to include the health and safety plan in a substantive draft form to the tendering contractors for their information.

The invitation to tender should request the tenderers to submit information to enable the client, or other appointer, to assess competence and the allocation of adequate resources. In particular, it is recommended that the tenderers submit a programme illustrating how time has been allocated for the management of health and safety.

Standard conditions of contract

Since the first edition of this book most, if not all, of the organisations responsible for publishing their own standard conditions of contracts now incorporate printed amendments to incorporate the requirements of the Regulations.

JCT Standard Form of Building Contract 1998 Edition

The Articles of Agreement for the traditional form of procurement identifies the planning supervisor at Article 6.1. The standard form envisages that the Architect will fulfil the role, although this can be easily amended to appoint any other party to the role.

The Appendix sets out the options for the Regulations. Either all the Regulations apply or only regulations 7 and 13 of the Regulations apply. It is not inconceivable that there may be situations when the Regulations do not apply at all, in which case the Fifth Recital would also be deleted.

Article 6.2 confirms that the principal contractor shall mean the Contractor. Again, it is not absolutely necessary for this to be the case and the Employer can appoint another contractor in the contractor's place.

Clause 6A of the Conditions sets out the provisions for use where the Appendix states that all the Regulations apply.

The Employer has a contractual obligation to ensure that the planning supervisor carries out all the duties of a planning supervisor under the Regulations. Thus, if the planning supervisor was to refuse to give advice to the Contractor pursuant to regulation 14(c)(i), the Employer would be liable for the costs incurred by the Contractor in respect of any expenditure that was necessary.

The Contractor also ensures, where the Contractor is not the principal contractor, that the principal contractor carries out all the duties of a principal contractor under the Regulations. The liability that the Employer has to the contractor under those circumstances could be exceptional and, for that reason, appointing another contractor other than the Contractor is unlikely to be commonplace, and then only undertaken in the light of careful advice.

The Contractor, in his capacity as the principal contractor, is obliged contractually to comply with all the duties of a principal contractor set out in the Regulations. Emphasis is given to the obligation to ensure that the health and safety plan complies with regulation 15(4). Any amendment by the Contractor to the health and safety plan has to be notified to the Employer who, in turn, is then required to notify the planning supervisor and the Architect. This has the benefit of maintaining the contractual relationships but, once again, imposes an obligation on the Employer that can produce adverse consequences if the contractual obligation to make such notifications is not effected.

If the Employer appoints a successor to the Contractor as the principal contractor, the Contractor has to comply with all the reasonable requirements of the new principal contractor. The Contractor is not entitled to any costs it may incur in complying with those reasonable requirements. Thus, costs associated in collating information for the health and safety plan would not be recoverable. The Contractor is also not entitled to any extension of time in complying with such a request.

The Contractor is required to provide information to the planning supervisor, within a reasonable time of being requested in writing by the planning supervisor, to provide information for the health and safety file.

ICE Conditions of Contract (Measurement Version) 7th Edition

The incorporation of the Regulations is set out, almost as an afterthought, in clause 71. The conditions of contract deal with health and safety issues, as indeed past editions have, in clauses 8(3), 15(2) and 19. By virtue of the 'standard' clauses, the Contractor has full responsibility for the adequacy, stability and safety of all site operations and methods of construction as set out in clause 8(3). This obligation is further reinforced by clause 15(2) that emphasises that the contractor is responsible for the safety of all operations. Just in case there was any doubt, the contractor is also to have full regard for the safety of all persons on site, as set out in clause 19(1). Although, clause 71(2) envisages that the Contractor is appointed principal contractor and the Engineer is appointed planning supervisor, there is some difficulty if the contractor is not the principal contractor. In such circumstances, the contract would have to identify the principal contractor in the special conditions and there would have to be clarification as to the responsibility for health and safety while noting that the principal contractor cannot contract out of its statutory obligations.

Since the Engineer does not have any express contractual obligations with respect to health and safety in the conditions of contract, the Engineer is fixed with express health and safety obligations as a result of the Regulations.

Clause 71(3)(a) provides that any action under the Regulations taken by either the planning supervisor or the Contractor under the Regulations shall be deemed to be an Engineer's instruction. This produces the absurd result of the Engineer instructing itself in the role of planning supervisor. Particular attention is drawn to any alteration or amendment to the health and safety plan that would require an Engineer's instruction. The Contractor is not entitled to any additional payment and/or extension of time if it is to any extent the fault of the Contractor.

The Engineer's key role in issuing instructions enhances responsibilities and, therefore, liabilities, as between the Engineer and the employer and the employer and the contractor.

If the action (as opposed to any lack of action or default) of either the planning supervisor or the principal contractor could not, in the Contractor's opinion, reasonably have been foreseen then the Contractor is required to give written notice as soon as possible to the Engineer. This raises the interesting possibility that the knowledge imputed to a principal contractor (who must be competent and have adequate resources) may well be different from the knowledge that can be imputed to an experienced contractor.

MF/1 (Revision 4) Model Form of General Conditions of Contract

Clause 20.1 makes the Contractor responsible for the adequacy, stability and safety of his operations on site. This standard form of contract will often involve works that do not come within the definition of construction work and, therefore, there is no express reference to planning supervisor or principal contractor and no contractual obligation to comply with the Regulations. It is advisable, in the event that the standard conditions of contract are being used for work that comes within the definition of construction work, that there are appropriate amendments to impose a contractual obligation on the parties to reflect the statutory obligations under the Regulations.

13 Liability and enforcement

Health and safety and the public debate

The public's perception and, therefore, confidence in the health and safety management of large companies is probably, at the time of writing, at an all time low. The Southall and Paddington rail disasters, and various other rail crashes, the Herald of Free Enterprise disaster, the King's Cross fire and the Marchioness sinking on the Thames all received wide press coverage and stimulated a debate that has created a powerful lobby to employ the full force of the criminal law against companies and management executives where death or injury is caused by negligence.

The Labour Government responded to the public's sense of outrage that there should be greater accountability for health and safety, by introducing a Bill that sought to create the offence of corporate manslaughter. Both companies and individuals, who contribute to the management of a company, would have been subject to a lower test of gross carelessness as compared with the current test of gross negligence. However, the Bill was withdrawn by the Government during 2000. The message is clear enough: health and safety is a management responsibility and criminal proceedings are a significant factor in reminding employees of the importance attached to carrying out their duties under health and safety laws. Further attempts to create new offences similar to the abandoned Bill can be expected in the future.

Enforcement

The Regulations will be enforced by the Executive in accordance with regulation 22 which provides:

> *Notwithstanding regulation 3 of the Health and Safety (Enforcing Authority) Regulations 1989, the enforcing authority for these Regulations shall be the Executive.*

This regulation should be read in conjunction with regulation 3(4), which provides:

> *The Regulations shall not apply to or in relation to construction work in respect of which the local authority within the meaning of regulation 2(1) of the Health and Safety (Enforcing Authority) Regulations 1989 is the enforcing authority.*

Thus in all cases, except where the Regulations do not apply (see Chapter 3), the Executive is the only enforcing authority.

The Executive's inspectors have wide ranging powers enabling them, for example, to require the production of documents, inspection of and have copies taken of such documents. The full extent of an inspector's powers are set out in section 20 of HSWA 1974. The Executive's enforcement powers if a statutory provision has been contravened include the service of an improvement notice (requiring a contravention to be remedied within a certain time period) or a prohibition notice (requiring that the activity in question be stopped until the contravention is remedied).

It may reasonably be assumed, and is borne out by experience, that the Executive's inspectors will require to see the health and safety plan as one of the first actions during a visit to a construction site. The inspector will be anxious to see that the management system on site has implemented the intentions developed in the health and safety plan; but, the outstanding difference in the manner in which the Regulations will be enforced, as compared with the practice prior to their implementation, will be visits to the offices of the clients, designers and planning supervisors.

During visits to the offices of clients, the inspector may be concerned to see how the client has satisfied himself with regard to the competence of the planning supervisor or principal contractor, or the decision making process in allowing construction work to commence. Visits to the offices of planning supervisors and designers may lead, typically, to lines of enquiry which will reveal whether the resources applied to the project are adequate, whether they have undertaken an appropriate risk assessment and whether they are co-operating and corresponding with other parties.

Table 1   Enforcement notices for the Regulations

Year	Improvement	Deferred prohibition	Immediate prohibition	Total
1995/96	19	2	44	65
1996/97	28	5	55	88
1997/98	23	7	88	118
1998/99	116	12	60	188
1999/00	90	7	64	161
Total	276	33	311	620

Enforcement notices

The number of enforcement notices issued by the Executive under the Regulations for the first five years is shown in Table 1.

Criminal proceedings

Despite the powers of enforcement, the criminal law is the principal instrument for securing compliance with the duties imposed by HSWA 1974 and all regulations made under it, including the Regulations. Section 33 of HSWA lists 15 separate offences which includes the contravention of any health and safety regulations or any requirement or prohibition imposed under any such regulations. The parties to a project are also liable inter alia to a criminal prosecution if they fail to comply with the requirements of the Executive's inspectors, which might, for example, involve the production of the health and safety plan and file and/or the design documentation. Most importantly, directors, managers and other officers of corporate organisations can be prosecuted personally where it is shown they have consented to, or connived at, the commission of any offence or where an offence has been committed by neglect on their part. The Executive has shown in recent years a greater willingness to prosecute individuals in addition to companies and other organisations. There can be no doubt that the trend of prosecuting individuals will continue.

The Magistrates' Courts have jurisdiction, according to the nature of the offence, to impose fines not exceeding £20,000 and a term of up to six months imprisonment, or both. Alternatively, if the offence is sufficiently serious to be tried in the Crown Court on indictment, the maximum fine is unlimited and imprisonment for a term not exceeding two years can be imposed, or both. In extreme cases a fine might be imposed in addition to a term of imprisonment.

Guidelines on sentencing

Although there has been a trend to impose larger fines for very serious offences, e.g. Heathrow Tunnel's collapse and the Southall rail disaster, there still remained a criticism that Courts were being too lenient with employers when prosecutions were brought under HSWA 1974 for work-based injuries. The Court of Appeal in *R. -v- F. Howe and Son (Engineers) Limited* took the opportunity of identifying guidelines on sentencing. The court stressed that every case would have to be decided on its own facts and avoided laying down any tariff or rule relating to the size of the fine to turnover or net profit. Nonetheless, in general terms, the following criteria were said to be relevant to sentencing:

(a)   How far short of the appropriate standard was the defendant's conduct in failing to reach the reasonably practical test?

(b)   Death is to be treated as an aggravating feature of an offence and public disquiet should be reflected in the penalty.

(c)   What was the degree of risk and the extent of the danger created?

(d)   What was the extent of the breach/breaches, e.g. an isolated incident or continuous over a period?

(e)   What are the defendant's resources and what effect will a fine have on its business?

(f)   Failure to heed warnings will be a particularly aggravating feature.

(g)   Any financial profit that the defendant has deliberately made from failing to take the necessary health and safety steps, or being prepared to run the risk of not taking them to save money, will be a particularly aggravating feature.

Mitigating factors that a Court will take into account will include:

- prompt admission of liability;

- timely plea of guilty;

- steps taken to remedy deficiencies after the Defendant became aware of them; and

- the defendant's good safety record.

In the later Court of Appeal case of *R. -v- Friskies Petcare Limited* it was recommended, based on the Howe guidelines, that the prosecution should present to the court any aggravating features in addition to the facts.

Under the Company Directors Disqualification Act 1986 a person can be disqualified from holding the office of director of a company after conviction of an indictable offence connected with the management of a company. This may extend to the management of health and safety matters. This extension of the concept of the management of a company was supported by Viscount Ullswater speaking for the Government in the House of Lords who said:

> *...in our view, Section 2 of the Company Directors Disqualification Act 1986 is capable of applying to health and safety matters. That Act provides for the court to make a disqualification order against a person connected with the promotion, formation, management or liquidation of a company. We believe that the potential scope of section 2(1) of that Act is very broad and that 'management' includes the management of health and safety.' (Hansard 28/11/91 Col. 1429)*

There has been at least one case where a company director has been disqualified. Mr Chapman was prosecuted by the Health and Safety Executive under section 37 of HSWA 1974. This enables proceedings to be taken against the director of a company which, with the director's consent, connivance or due to his negligence, committed an offence. In this case, Mr Chapman's company, Chapman Chalk Supplies Limited, had contravened a prohibition notice. Mr Chapman was fined £5,000 and a separate fine for the same amount was imposed on the company. The court also used its powers under section 2(1) of the Company Directors Disqualification Act 1986 to ban Mr Chapman from being a company director for two years (G. Slapper, 'Where the Buck Stops', *New Law Journal*, 24 July 1992).

Note that a conviction under HSWA 1974, or for any other crime, by virtue of section 11 of the Civil Evidence Act 1968 is admissible as evidence in any subsequent civil proceedings, subject to such evidence being relevant to any issue in those proceedings. In civil proceedings, discussed below, a conviction constitutes the basic fact of a presumption. The defendant would have the uphill task of persuading the court that the verdict beyond reasonable doubt was wrong.

Failure to comply with the ACOP is not in itself an offence, although it may be taken as proof that a person has contravened the relevant regulation (Section 17 – Health and Safety at Work etc. Act 1974). It would, however, be open to the accused to prove that compliance with the Regulations had been achieved in some other way.

Comment and analysis of prosecutions under the Regulations

There have been 167 prosecutions up to end of 1999/2000 of which 110 have been successful. Capturing details of the prosecutions is not precise because the majority of prosecutions are in the Magistrates Courts and, therefore, are not reported. It is often the case that companies or individuals that have been prosecuted under the Regulations have also been charged with associated offences under HSWA 1974 and other regulations. However, it is possible to identify some strong trends linking the incidents, or likelihood, or prosecution with certain regulations, see Table 2.

Table 2   Relationship between the likelihood of prosecution and the Regulations

High incidence of prosecutions
reg. 6(1)
reg. 10
Medium incidence of prosecutions
reg. 16(1)
reg. 15(4)
reg. 11
Low incidence of prosecution
reg. 8(3)
reg. 15(1)
reg. 6(3)
reg. 7(5)
reg. 13
reg. 14
reg. 19(1)

The Executive now update a website that contains brief details of all prosecutions under health and safety legislation in the construction industry. The website address is www.hse-databases.co.uk/prosecutions.

There have been no prosecutions, to my knowledge, under any of the Regulations not listed in Table 2.

A feature shared by nearly all the prosecutions is that they are usually connected with accidents or other incidents, particularly where children and public safety are at issue. In other words, the Executive are not being proactive in prosecuting as a deterrent but are being reactive to what has already manifested itself as a failure in the management of health and safety.

It is also interesting to note that the role that is most often involved in prosecutions is the client. There is either a failure to appoint a planning supervisor and/or a principal contractor which one would have associated, in many cases, with a failure by the designer to inform the client of its obligations, as required by regulation 13(1). However, this is not the case as there has only been one prosecution under regulation 13. The failure to ensure that the health and safety plan has been prepared and approved before construction work commences is also a common basis for the prosecution because, in many cases, there are still construction projects without health and safety plans. Whether or not the Executive have chosen to target clients, it indicates that the role of the client in influencing health and safety management is to the fore and has not been given sufficient priority by clients to date.

There has been a recent trend that principal contractors have been prosecuted at an increasing rate. This is in stark contrast to the very low numbers of prosecutions of planning supervisors and designers. The reason for this may be simply that it is more difficult to prove the breach of obligations which are, by their nature, more subjective than the obvious failures that are associated with contractors.

The chief inspector for construction has said that the Executive will be targeting designers, not least because it is recognised that designers have a profound influence on the health and safety management of construction projects. It will be interesting to observe whether or not the Executive is able to maintain successful prosecutions against designers in the future.

The largest fine under the Regulations has been £50,000 in the Crown Court. The majority of fines in the Magistrates Court lie in the range £100 to £5,000. The average fine is approximately £2,000 for breaches of each regulation.

It is also interesting to note that the Executive have prosecuted individuals as well as limited companies, although the level of fines for individuals has been very much at the lower end of the scale.

Civil liability

In addition to the liability of a criminal prosecution, persons in breach of the Regulations may be liable in civil proceedings in certain circumstances.

Section 47(2) of HSWA 1974 provides:

> *Breach of duty imposed by health and safety Regulations shall, so far as it causes damage, be actionable except insofar as the Regulations provide otherwise.*

In the context of HSWA 1974 and the regulations made further to Article 118A of the Treaty of Rome, which seek improvements in occupational health and safety in an extensive and far reaching programme of legislation, it is difficult to understand why civil liability is excluded specifically from selected regulations.

In particular the Framework Regulations, which have to be read with all the other regulations, implementing the daughter directives of the Framework Directive, provide:

> *Breach of a duty imposed by these Regulations shall not confer a right of action in any civil proceedings.*

Thus, a breach of any of the general duties in the Framework Regulations cannot be the basis for an action of breach of statutory duty. However, the 'daughter' directives and the other five regulations which make up the 'six-pack' do not exclude civil liability whereas, still more confusingly, the Regulations do exclude civil liability, except in respect of regulation 10 and regulation 16(1)(c) as provided by regulation 21:

> *Breach of a duty imposed by these Regulations other than those imposed by regulation 10 and regulation 16(1)(c), shall not confer a right of action in any civil proceedings.*

Where does all this leave the would-be litigant seeking a remedy in civil law under the Regulations?

In a system that relies on proving 'fault' to obtain damages or compensation for personal injuries or, in the case of bereaved dependants, death, a litigant has to frame the case within a recognised cause of action. The causes of action which are commonly pursued in personal injury claims arising from occupational accidents are breach of statutory duty and common law negligence.

Breach of statutory duty

Breach of statutory duty is available as a cause of action in respect of regulation 10 and regulation 16(1)(c). The litigant would have to prove a number of requirements which are common to any claim for breach of statutory duty.

A duty imposed on the defendant himself

The Plaintiff must prove, which should not be difficult, that a mandatory duty is imposed on the Defendant by the Regulations if the action is to proceed.

The duty was owed to that plaintiff

The classes of persons who might be able to rely on regulations 10 and 16(1)(c) are not limited and would include employees of the parties to a project and members of the public at large.

The duty was owed by that defendant

The Regulations have defined the roles of client, planning supervisor, designer, principal contractor and contractor, which should make it relatively straightforward to fix a Defendant, fulfilling one of the roles, with the relevant duty.

The defendant was in breach of the statutory duty

(i)   *Regulation 10*

A client has to ensure, in compliance with regulation 10, that a health and safety plan has been prepared in respect of a project before the start of the construction phase. However, the duty is tempered by the qualification that a client shall only ensure the health and safety plan has been prepared 'so far as reasonably practicable'. Thus, it is always open for a client to prove that the steps or measures taken to comply with regulation 10 were 'reasonably practicable'.

The words, 'so far as reasonably practicable' are used frequently in the 'six-pack' regulations and the Regulations but without any guidance as to how far or what steps or measures a client should take. The ACOP recognises that a client has to make a judgment as to whether the health and safety plan complies with regulation 15(5). There is no guidance as to whether a client's judgment should be assessed against an objective or subjective criteria, although a client is able to take the advice of the planning supervisor or others, which suggests a client's judgment would be considered on an objective basis. But what if the planning supervisor providing the advice is not competent? This can give rise to a potential claim in negligence as discussed below.

The guidance provided by the ACOP fails to address the test of whether the steps or measures to be taken by a client are reasonably practicable. The suggestion is offered that it would be unreasonable to assess the risk to health against the steps or measures which would have to be taken to ensure absolute compliance with regulation 10. The likelihood is that, if the steps or measures to be taken by the client are grossly disproportionate to the risk the measures may exceed what is 'reasonably practicable'. It is difficult to imagine any steps or measures which a client would need to take to comply with regulation 10 which would be so unreasonable as to provide any defence. This must be so because a client has the contractual authority over a contractor to instruct the stopping or starting of the construction phase.

(ii) *Regulation 16(1)(c)*

The principal contractor is exposed to an action for breach of statutory duty in regulation 16(1)(c). The principal contractor is required to exclude unauthorised persons from any premises or part of premises where construction is being carried out. The steps which the principal contractor is required to take have only to be 'reasonable'. The rationale as to what is, or is not, reasonable as applied to a client and referred to above under regulation 10, applies equally to regulation 16(1)(c). The balance between risk to the health and safety of unauthorised persons on premises where construction work is being carried out and the steps to ensure only authorised persons can enter into such premises is more difficult to strike than in the case of a client under regulation 10. The ACOP acknowledges that exclusion measures must be related to what is foreseeable. The ACOP draws the comparison between the example of a large remote site and a site close to a school where exclusion measures in the latter case would be expected to be more effective than in the former case. Despite the provision of security staff, high fences and other forms of deterrent, all of which have an easily identifiable cost, there is no system of preventing unauthorised persons entering into premises which is foolproof.

Note that the ACOP will be admissible as evidence in civil actions with the advantage of establishing for a plaintiff the nature of the duties expected of a client and principal contractor.

Nothing in the Regulations detracts from the principal contractor's, or contractor's, obligations to visitors and/or trespassers as provided for by the Occupier's Liability Acts of 1957 and 1984 respectively.

The harm suffered by the plaintiff was of the type which the regulations sought to prevent

The rule in *Gorris -v- Scott 1874* has led to a strict and narrow interpretation of statutory requirements. In the interpretation of Section 14 of the Factories Act 1961 the House of Lords held that the statutory duty on an employer to fence every dangerous part of a machine was confined to keeping the worker out of the machine, not to protect him from injury caused by parts of the machine being ejected. Despite this principle having been criticised widely in later cases, this approach to interpreting statutory requirements has not changed. However, the overlap between all the recent health and safety regulations should improve the chances that prevention of a particular harm is the purpose of at least one regulation.

The harm which could be interpreted to be prevented by regulation 10 is extremely wide. It is possible that any person suffering any harm could argue that the harm was of a kind which the regulation sought to prevent. Equally, the harm which regulation 16(1)(c) seeks to prevent would be harm associated with trespass and interference with plant and machinery.

There is no certainty that an action for breach of statutory duty under regulations 10 and 16(1)(c) would succeed for the reasons alluded to and explained above. The opportunity to pursue causes of action for breach of statutory duty under other regulations may also exist despite the apparent exclusion in regulation 21. It is implicit in regulation 10 that the client is satisfied the health and safety plan is adequate and complies with regulation 15. Since the health and safety plan is one of the central innovations of the Regulations if it can be proved that a client has no grounds for being satisfied that it complies with regulation 15(4), a cause of action would exist for breach of regulation 10. By a side wind, as it were, other requirements, particularly with regard to the competence of the planning supervisor and principal contractor and compliance with regulation 8, could be used as evidence of failure to comply with regulation 10.

Finally, if a plaintiff has been deprived of a civil remedy for breach of a statutory duty, the lack of such a remedy might be corrected by arguing there is a failure to implement fully, the Temporary or Mobile Worksites Directive because the Regulations fail to provide remedies which are realistic (*Von Colson -v- Land Nordrhein-Westfalen*).

Negligence

The express exclusion of a cause of action based on breach of statutory duty does not exclude a common law claim for negligence. The essential elements of negligence are:

1.   A duty owed by the defendant to the plaintiff;

2.   Breach of the duty owed by the defendant;

3.   Breach of the duty caused damage.

The duty owed by an employer to its employees was summed up in the leading case of *Wilsons and Clyde Coal Co. Ltd -v- English 1938* as:

> *A duty which rests on the employer and which is personal to the employer, to take reasonable care for the safety of his workmen, whether the employer be an individual, a firm, or a company, and whether or not the employer takes any share in the conduct of the operations.*

The duty owed by employers has been categorised by the courts as a duty to provide:

1.   A safe workplace;

2.   Safe equipment;

3.   Competent staff and fellow workmen;

4.   A safe system of work.

The entire philosophy behind the Regulations is to provide a safer workplace on site, together with a safer system of work, or method of construction, and that the planning supervisor, designers and the principal contractor are required to be competent. There will be overlap with other regulations, particularly the Work Equipment Regulations in respect of the requirement to provide safe equipment and the Framework Regulations which deal with general principles of risk assessment.

Although the Regulations do not themselves support civil liability, except for regulations 10 and 16(1)(c), they are evidence of required practice and failure to follow such practice, such as the appointment of a competent designer, can constitute negligence.

Contributory negligence

Many personal injury claims arising from industrial accidents, whether based on breach of statutory duty or negligence, often have to overcome an allegation of contributory negligence. Contributory negligence is proved when:

(i)   The injury of which the plaintiff complains results from that particular risk to which the negligence of the plaintiff exposed him;

(ii) The negligence of the plaintiff contributed to his injury; and

(iii) There was fault or negligence on the part of the plaintiff.

The extent to which an employee has suffered as a result of contributory negligence will depend on the facts in each case.

Section 7 of HSWA 1974 provides:

It shall be the duty of every employee while at work –

(a)   *to take reasonable care for the health and safety of himself and of other persons who may be affected by his acts or omissions at work; and*

(b)   *as regards any duty or requirement imposed on his employer or any other person by or under any of the relevant statutory provisions, to co-operate with him so far as is necessary to enable that duty or requirement to be performed or complied with.*

The 'six-pack' regulations also impose duties on employees which provide greater scope for employers to prove contributory negligence. A discussion on contributory negligence of employees and the statutory duties which they owe to their employers is beyond the scope of this book but is mentioned for the sake of completeness.

14 Transitional provisions

The Regulations came into force on 31 March 1995, however the impact of the Regulations on the construction industry was such that the industry could not be reasonably expected to comply fully with all the obligations and duties from the moment they came into force. The transitional provisions have expired such that all projects are now subject to the Regulations. However, the impact of the transitional provisions may not be without some force with respect to disputes, which includes some aspect of the Regulations, arising before 1 January 1996.

The transitional provisions are referred to in regulation 23, which provides:

> *Schedule 2 shall have effect with respect to projects which have started, but the construction phase of which has not ended, when these Regulations come into force.*

Therefore, the transitional provisions applied to all projects, to a limited extent, provided that the construction phase of a project had not ended on or before 30 March 1995.

Paragraph 1 of schedule 2 provides:

> *Until 1st January 1996 regulation 6 shall not apply in respect of a project the construction phase of which started before the coming into force of these Regulations.*

Regulation 6 is concerned with the appointment of the planning supervisor and principal contractor. Therefore, clients were not obliged to appoint a planning supervisor or principal contractor if the construction phase had started before 31 March 1995. However, if the construction phase was likely to extend beyond 1 January 1996, the client was obliged to make the appropriate appointments of the planning supervisor and principal contractor not later than 31 December 1995.

Paragraph 2 of schedule 2 provides:

> *Where at the coming into force of these Regulations the time specified in regulation 6(3) for the appointment of the planning*

supervisor has passed, the time for appointing the planning super-visor by virtue of regulation 6(1)(a) shall be as soon as is practica-ble after the coming into force of these Regulations.

If the design or construction phase of a project had not started before 31 March 1995 a client had to appoint a planning supervisor pursuant to regulation 6(1)(a), in any event, and as soon as possible after the coming into force of the Regulations. A client would normally have had to appoint a planning supervisor as soon as possible after collecting enough informa-tion about the project and the construction work involved in it (which would usually be before the appointment of designers). The effect of the transitional provisions was to oblige a client to appoint a planning supervisor as soon as practicable even if construction work was to have started as early as 1 April 1995.

Paragraph 3 of schedule 2 provides:

Where at the coming into force of these Regulations the time specified in regulation 6(4) for the appointment of the principal contractor has passed, the time for appointing the principal contractor by virtue of regulation 6(1)(b) shall be as soon as is practicable after the coming into force of these Regulations.

The requirements on a client to appoint a principal contractor are similar to the requirements under paragraph 2 of schedule 2. Therefore, a client was obliged to appoint a principal contractor even if construction work was to start as early as 1 April 1995. A contractor appointed as principal contractor had the difficult task of developing the health and safety plan after the planning supervisor had ensured a plan was prepared and handed over to the principal contractor.

Paragraph 4 of schedule 2 provides:

Regulation 7 shall not require notification of any project where notice of all construction work included in the project has been given in accordance with section 127(6) of the Factories Act 1961 before the coming into force of these Regulations.

Therefore if notice for a project had been given in accordance with section 127(6) of the Factories Act 1961 on or before 30 March 1995 the require-ments under regulation 7 do not apply – but only if the notice included all construction work.

Paragraph 5 of schedule 2 provides:

Regulation 10 shall not apply to any project the construction phase of which starts before 1st August 1995.

A client did not have to comply with regulation 10 if the construction phase of a project started at any time before 1 August 1995, which meant that such a project did not need a health and safety plan complying with regulation 15(4) for the purposes of satisfying the client, before the start of the construction phase. However, that did not mean a health and safety plan should not be prepared. Thus, for a project starting on 1 April 1995 the client was obliged to appoint a planning supervisor and principal contractor who had, in turn, the duty to prepare and develop the health and safety plan. However, they did not have to produce the health and safety plan for the client's satisfaction as complying with regulation 15(4), before the construction phase commenced. Note that if the design was completed before 1 August 1995, the health and safety plan did not have to include the design information set out in regulations 13(2)(a) and 13(2)(b); see paragraph 7 of schedule 2 below.

Paragraph 6 of schedule 2 provides:

Regulation 11 shall not apply to any project the construction phase of which started before the coming into force of these Regulations.

In the case of the construction phase for a project which had started on 31 March 1995, or later, a client had to ensure the planning supervisor had the information referred to in regulation 11. This was necessary because the health and safety plan was still required. That produced the odd result that in the case of a project where the construction phase started before 31 March 1995 and a planning supervisor and principal contractor were appointed in accordance with paragraphs 2 and 3 of schedule 2 the health and safety plan did not have to contain the information which a client might have produced by virtue of regulation 11.

Paragraph 7 of schedule 2 provides:

Until 1st August 1995 regulation 13 and regulation 14(a) shall not apply in respect of any design the preparation of which started before the coming into force of these Regulations.

Designers escaped the requirements of regulation 13 up to 1st August 1995 if the design had started before 31st March 1995. Note that even if the design was completed before 1st August 1995 the designer was obliged to co-operate with the planning supervisor and any other designer to enable them to comply with the requirements and prohibitions placed on

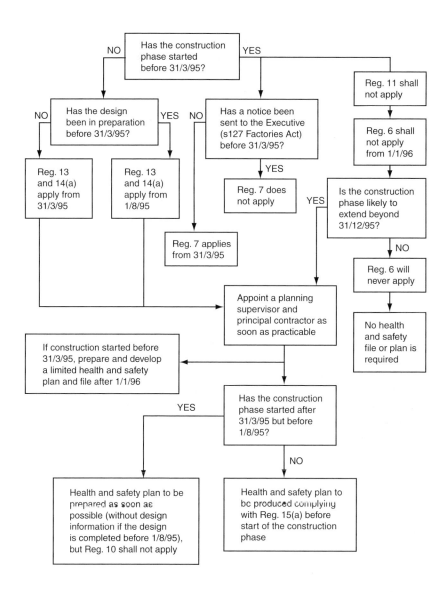

Figure 6 Flowchart illustrating the transitional provisions in Schedule 2 of the Regulations

each of them. Equally, the requirements on the planning supervisor to ensure the design included the needs and information set out in regulations 13(2)(a) and 13(2)(b) did not apply until 1st August 1995. Accordingly, if the designer had completed the design before 1st August 1995 this requirement had no effect and the health and safety plan which the planning supervisor was still required to prepare did not necessarily include such design information.

The transitional provisions were complicated and the flowchart in Figure 6 will be of assistance in determining to what extent the Regulations applied according to the stage of the project.

Bibliography

Baumert K., Kete N. and Figueres C. *CDM Design: How an "Open Architecture" Can Meet the Needs and Visions*. World Resources Institute.

Bennett L. (ed.) (2000). *Adding Value Through the Project Management of CDM*. Thomas Telford Publishing, London.

Building Performance Services Ltd (2001). *Understanding the CDM Regulations*. Spon Press, Taylor & Francis Books Ltd, London.

Construction Industry Council (1995). *Construction Industry Council Digest Issue 17, Construction (Design and Management) Regulations Special*. 18 January 1995.

Construction Industry Research and Information Association (CIRIA) (1997). *Experiences of CDM*. CIRIA, London.

Construction Industry Research and Information Association (CIRIA) (1998). *CDM Regulations – Practical Guidance for Clients and Clients' Agents*. CIRIA, London.

Cooks J., Briffa G., Dew L., Watson D., Stokes M. *et al.* (1995). *CDM Regulations*. CIRIA, London.

Croner Publications (1994). *Croner's Health & Safety at Work* (updated November 1994). Croner Publications Ltd.

Department of the Environment (1994). Press Release, 25 July.

Hansard (28/11/91 Col 1429).

Health and Safety Executive (HSE) (1988). *Blackspot Construction*. HSE.

Health and Safety Executive (HSE) (1995). *Managing Construction for Health and Safety: Construction (Design and Management) Regulations 1994*, Approved Code of Practice. HSE.

Health and Safety Executive (HSE) (1997). *Evaluation of the CDM Regulations 1994*. HSE.

Health and Safety Information Bulletin (June 1987).

Institution of Civil Engineers (ICE) (1999). *ICE Conditions of Contract Measurement Version*, Seventh Edition. Thomas Telford Publishing, London.

JCT Standard Form of Building Contract, 1998 Edition.

Latham M. Sir (1994). *Constructing the Team*. Report July 1994. Her Majesty's Stationery Office, London.

Nanayakkara R. (1997). *CDM Regulations – Health and Safety File*. BSRIA Ltd.

Nanayakkara R. (1997). *Standard Specification for the CDM Regulations Health and Safety File*. BSRIA Ltd.

Ove Arup & Partners (1997). *CDM Regulations*. Construction Industry Research and Information Association, London.

Oxenburgh M. (1991). *Increasing Productivity and Profit Through Health and Safety*. CCH Editions.

Perry P. (1999). *CDM Questions and Answers*. Thomas Telford Publishing, London.

PowerGen plc. (1996). *The CDM Regulations*. Blackwell Science Ltd.

Royal Institute of British Architects (RIBA) (1985). *Engaging an Architect: Guidance for Client on CDM Regs*. RIBA Publications Ltd, London.

Royal Institute of British Architects (RIBA) (1996). *Architect's Guide to Job Administration Under the CDM Regulations*. RIBA Publications Ltd, London.

Slapper G. (1992). Where the Buck Stops. *New Law Journal*. 24 July.

Smith I., Goddard C. and Randall N. (1993). *Health and Safety – A New Legal Framework*. Butterworth.

Spiers A. R., RSP-FIOSH. (1998). *CDM Regulations 1994*. The College of Estate Management.

Summerhayes S. and Williams T. (1999). *The CDM Regulations Procedures Manual*. Blackwell Science Ltd.

Thorpe B. (1999). *Addressing CDM Regulations Through Quality Management Systems*. Gower Publishing Limited.

Appendix 1

COUNCIL DIRECTIVE 92/57/EEC OF 24 JUNE 1992
on the implementation of minimum safety and health requirements at temporary or mobile constructions sites (eighth individual Directive within the meaning of Article 16 (1) of Directive 89/391/EEC)

THE COUNCIL OF THE EUROPEAN COMMUNITIES,
Having regard to the Treaty establishing the European Economic Community, and in particular Article 118a thereof,

Having regard to the proposal from the Commission, submitted after consulting the Advisory Committee on Safety, Hygiene and Health Protection at Work,

In cooperation with the European Parliament,

Having regard to the opinion of the Economic and Social Committee,

Whereas Article 118a of the Treaty provides that the Council shall adopt, by means of directives, minimum requirements for encouraging improvements, especially in the working environment, to ensure a better level of protection of the safety and health of workers;

Whereas, under the terms of that Article, those directives are to avoid imposing administrative, financial and legal constraints in a way which would hold back the creation and development of small and medium-sized undertakings;

Whereas the communication from the Commission on its programme concerning safety, hygiene and health at work provides for the adoption of a Directive designed to guarantee the safety and health of workers at temporary or mobile construction sites;

Whereas, in its resolution of 21 December 1987 on safety, hygiene and health at work, the Council took note of the Commission's intention of submitting to the Council in the near future minimum requirements concerning temporary or mobile construction sites;

Whereas temporary or mobile construction sites constitute an area of activity that exposes workers to particularly high levels of risk;

Whereas unsatisfactory architectural and/or organizational options or poor planning of the works at the project preparation stage have played a role in more than half of the occupational accidents occurring on construction sites in the Community;

Whereas in each Member State the authorities responsible for safety and health at work must be informed, before the beginning of the works, of the execution of works the scale of which exceeds a certain threshold;

Whereas, when a project is being carried out, a large number of occupational accidents may be caused by inadequate coordination, particularly where various undertakings work simultaneously or in succession at the same temporary or mobile construction site;

Whereas it is therefore necessary to improve coordination between the various parties concerned at the project preparation stage and also when the work is being carried out;

Whereas compliance with the minimum requirements designed to guarantee a better standard of safety and health at temporary or mobile construction sites is essential to ensure the safety and health of workers;

Whereas, moreover, self-employed persons and employers, where they are personally engaged in work activity, may, through their activities on a temporary or mobile construction site, jeopardize the safety and health of workers;

Whereas it is therefore necessary to extend to self-employed persons and to employers where they are personally engaged in work activity on the site certain relevant provisions of Council Directive 89/655/EEC of 30 November 1989 concerning the minimum safety and health requirements for the use of work equipment by workers at work (second individual Directive) (6), and of Council Directive 89/656/EEC of 30 November 1989 on the minimum health and safety requirements for the use by workers of personal protective equipment at the workplace (third individual Directive);

Whereas this Directive is an individual Directive within the meaning of Article 16 (1) of Council Directive 89/391/EEC of 12 June 1989 on the introduction of measures to encourage improvements in the safety and health of workers at work; whereas, therefore, the provisions of the said Directive are fully applicable to temporary or mobile construction sites, without prejudice to more stringent and/or specific provisions contained in this Directive;

Whereas this Directive constitutes a practical step towards the achievement of the social dimension of the internal market with special reference to the subject matter of Council Directive 89/106/EEC of 21 December 1988 on the approximation of laws, regulations and administrative provisions of the Member States relating to construction products and the subject matter covered by Council Directive 89/440/EEC of 18 July 1989 amending Directive 71/305/EEC concerning coordination of procedures for the award of public work contracts;

Whereas, pursuant to Council Decision 74/325/EEC the Advisory Committee on Safety, Hygiene and Health Protection at Work is consulted by the Commission with a view to drawing up proposals in this field,

HAS ADOPTED THIS DIRECTIVE:

Article 1
SUBJECT

1.   This Directive, which is the eighth individual Directive within the meaning of Article 16 (1) of Directive 89/391/EEC, lays down minimum safety and health requirements for temporary or mobile construction sites, as defined in Article 2 (a).

2.   This Directive shall not apply to drilling and extraction in the extractive industries within the meaning of Article 1 (2) of Council Decision 74/326/EEC of 27 June 1974 on the extension of the responsibilities of the Mines Safety and Health Commission to all mineral-extracting industries.

3.   The provisions of Directive 89/391/EEC are fully applicable to the whole scope referred to in paragraph 1, without prejudice to more stringent and/or specific provisions contained in this Directive.

Article 2
DEFINITIONS

For the purposes of this Directive:

(a) 'temporary or mobile construction sites' (herinafter referred to as 'construction sites') means any construction site at which building or civil engineering works are carried out; a non-exhaustive list of such works is given in Annex 1;

(b) 'client' means any natural or legal person for whom a project is carried out;

(c) 'project supervisor' means any natural or legal person responsible for the design and/or execution and/or supervision of the execution of a project, acting on behalf of the client;

(d) 'self-employed person' means any person other than those referred to in Article 3 (a) and (b) of Directive 89/391/EEC whose professional activity contributes to the completion of a project;

(e) 'coordinator for safety and health matters at the project preparations stage' means any natural or legal person entrusted by the client and/or project supervisor, during preparation of the project design, with performing the duties referred to in Article 5;

(f) 'coordinator for safety and health matters at the project execution stage' means any natural or legal person entrusted by the client and/or project supervisor, during execution of the project, with performing the duties referred to in Article 6.

Article 3

APPOINTMENT OF COORDINATORS – SAFETY AND HEALTH PLAN – PRIOR NOTICE

1.   The client or the project supervisor shall appoint one or more coordinators for safety and health matters, as defined in Article 2 (e) and (f), for any construction site on which more than one contractor is present.

2.   The client or the project supervisor shall ensure that prior to the setting up of a construction site a safety and health plan is drawn up in accordance with Article 5 (b).

The Member States may, after consulting both management and the workforce, allow derogations from the provisions of the first paragraph, except where the work concerned involves particular risks as listed in Annex II.

3.   In the case of constructions sites:

   –   on which work is scheduled to last longer than 30 working days and on which more than 20 workers are occupied simultaneously, or

- on which the volume of work is scheduled to exceed 500 person days,

the client or the project supervisor shall communicate a prior notice drawn up in accordance with Annex III to the competent authorities before work starts.

The prior notice must be clearly displayed on the construction site and, if necessary, periodically updated.

Article 4
PROJECT PREPARATION STAGE: GENERAL PRINCIPLES

The project supervisor, or where appropriate the client, shall take account of the general principles of prevention concerning safety and health referred to in Directive 89/391/EEC during the various stages of designing and preparing the project, in particular:

- when architectural, technical and/or organizational aspects are being decided, in order to plan the various items or stages of work which are to take place simultaneously or in succession,

- when estimating the period required for completing such work or work stages. Account shall also be taken, each time this appears necessary, of all safety and health plans and of files drawn up in accordance with Article 5 (b) or (c) or adjusted in accordance with Article 6 (c).

Article 5
PROJECT PREPARATION STAGE: DUTIES OF COORDINATORS

The coordinator(s) for safety and health matters during the project preparation stage appointed in accordance with Article 3 (1) shall:

(a)   coordinate implementation of the provisions of Article 4;

(b)   draw up, or cause to be drawn up, a safety and health plan setting out the rules applicable to the construction site concerned, taking into account where necessary the industrial activities taking place on the site; this plan must also include specific measures concerning work which falls within one or more of the categories of Annex II;

(c)   prepare a file appropriate to the characteristics of the project containing relevant safety and health information to be taken into account during any subsequent works.

Article 6

PROJECT EXECUTION STAGE: DUTIES OF COORDINATORS

The coordinator(s) for safety and health matters during the project execution stage appointed in accordance with Article 3 (1) shall:

(a) coordinate implementation of the general principles of prevention and safety:

–   when technical and/or organizational aspects are being decided, in order to plan the various items or stages of work which are to take place simultaneously or in succession,

–   when estimating the period required for completing such work or work stages;

(b) coordinate implementation of the relevant provisions in order to ensure that employers and, if necessary for the protection of workers, self-employed persons:

–   apply the principles referred to in Article 8 in a consistent manner,

–   where required, follow the safety and health plan referred to in Article 5 (b);

(c) make, or cause to be made, any adjustments required to the safety and health plan referred to in Article 5 (b) and the file referred to in Article 5 (c) to take account of the progress of the work and any changes which have occurred;

(d) organize cooperation between employers, including successive employers on the same site, coordination of their activities with a view to protecting workers and preventing accidents and occupational health hazards and reciprocal information as provided for in Article 6 (4) of Directive 89/391/EEC, ensuring that self-employed persons are brought into this process where necessary;

(e) coordinate arrangements to check that the working procedures are being implemented correctly;

(f) take the steps necessary to ensure that only authorized person are allowed onto the construction site.

Article 7

RESPONSIBILITIES OF CLIENTS, PROJECT SUPERVISORS AND EMPLOYERS

1.   Where a client or project supervisor has appointed a coordinator or coordinators to perform the duties referred to in Articles 5 and 6, this does not relieve the client or project supervisor of his responsibilities in that respect.

2.   The implementation of Articles 5 and 6, and of paragraph 1 of this Article shall not affect the principle of employers' responsibility as provided for in Directive 89/391/EEC.

Article 8

IMPLEMENTATION OF ARTICLE 6 OF DIRECTIVE 89/391/EEC

When the work is being carried out, the principles set out in Article 6 of Directive 89/391/EEC shall be applied, in particular as regards:

(a)   keeping the construction site in good order and in a satisfactory state of cleanliness;

(b)   choosing the location of workstations bearing in mind how access to these workplaces is obtained, and determining routes or areas for the passage and movement and equipment;

(c)   the conditions under which various materials are handled;

(d)   technical maintenance, pre-commissioning checks and regular checks on installations and equipment with a view to correcting any faults which might affect the safety and health of workers;

(e)   the demarcation and laying-out of areas for the storage of various materials, in particular where dangerous materials or substances are concerned;

(f)   the conditions under which the dangerous materials used are removed;

(g)   the storage and disposal or removal of waste and debris;

(h)   the adaptation, based on progress made with the site, of the actual period to be allocated for the various types of work or work stages;

(i)   cooperation between employers and self-employed persons;

(j)   interaction with industrial activities at the place within which or in the vicinity of which the construction site is located.

Article 9

OBLIGATIONS OF EMPLOYERS

In order to preserve safety and health on the construction site, under the conditions set out in Article 6 and 7, employers shall:

(a) in particular when implementing Article 8, take measures that are in line with the minimum requirements set out in Annex IV;

(b) take into account directions from the coordinator(s) for safety and health matters.

Article 10

OBLIGATIONS OF OTHER GROUPS OF PERSONS

1.   In order to preserve safety and health on the construction site, self-employed persons shall:

(a) comply in particular with the following, *mutatis mutandis*:

   (i)   the requirements of Article 6 (4) and Article 13 of Directive 89/391/EEC and Article 8 and Annex IV of this Directive;

   (ii) Article 4 of Directive 89/655/EEC and the relevant provisions of the Annex thereto;

   (iii) Article 3, Article 4 (1) to (4) and (9) and Article 5 of Directive 89/656/EEC;

(b) take into account directions from the coordinator(s) for safety and health matters.

2.   In order to preserve safety and health on the site, where employers are personally engaged in work activity on the construction site, they shall:

(a) comply in particular with the following, *mutatis mutandis*:

   (i)   Article 13 of Directive 89/391/EEC;

   (ii) Article 4 of Directive 89/655/EEC and the relevant provisions of the Annex thereto;

   (iii) Articles 3, 4 (1), (2), (3), (4), (9) and 5 of Directive 89/656/EEC;

(b) take account of the comments of the coordinator(s) for safety and health.

Article 11

INFORMATION FOR WORKERS

1.   Without prejudice to Article 10 of Directive 89/391/EEC, workers and/or their representatives shall be informed of all the measures to be taken concerning their safety and health on the construction site.

2.   The information must be comprehensible to the workers concerned.

Article 12

CONSULTATION AND PARTICIPATION OF WORKERS

Consultation and participation of workers and/or of their representatives shall take place in accordance with Article 11 of Directive 89/391/EEC on matters covered by Articles 6, 8 and 9 of this Directive, ensuring whenever necessary proper coordination between workers and/or workers' representatives in undertakings carrying out their activities at the workplace, having regard to the degree of risk and the size of the work site.

Article 13

AMENDMENT OF THE ANNEXES

1.   Amendments to Annexes I, II and III shall be adopted by the Council in accordance with the procedure laid down in Article 118a of the Treaty.

2.   Strictly technical adaptations of Annex IV as a result of:

   - the adoption of directives on technical harmonization and standardization regarding temporary or mobile construction sites, and/or

   - technical progress, changes in international regulations or specifications or knowledge in the field of temporary or mobile construction sites

shall be adopted in accordance with the procedure laid down in Article 17 of Directive 89/391/EEC.

Article 14

FINAL PROVISIONS

1.   Member States shall bring into force the laws, regulations and administrative provisions necessary to comply with this Directive by 31 December 1993 at the latest.

They shall forthwith inform the Commission thereof.

2.   When Member States adopt these measures, they shall contain a reference to this Directive or be accompanied by such reference on the occasion of their official publication. The methods of making such a reference shall be laid down by the Member States.

3.   Member States shall communicate to the Commission the texts of the provisions of national law which they have already adopted or adopt in the field governed by this Directive.

4.   Member States shall report to the Commission every four years on the practical implementation of the provisions of this Directive, indicating the points of view of employers and workers.

The Commission shall inform the European Parliament, the Council, the Economic and Social Committee and the Advisory Committee on Safety, Hygiene and Health Protection at Work.

5.   The Commission shall submit periodically to the European Parliament, the Council and the Economic and Social Committee a report on the implementation of this Directive, taking into account paragraphs 1, 2, 3 and 4.

Article 15

THIS DIRECTIVE IS ADDRESSED TO THE MEMBER STATES

Done at Luxembourg, 24 June 1992.

For the Council
The President
Jose da SILVA PENEDA

ANNEX I

NON-EXHAUSTIVE LIST OF BUILDING AND CIVIL ENGINEERING WORKS REFERRED TO IN ARTICLE 2 (a) OF THE DIRECTIVE

1.   Excavation

2.   Earthworks

3.   Construction

4.   Assembly and disassembly of prefabricated elements

5. Conversion or fitting-out

6. Alterations

7. Renovation

8. Repairs

9. Dismantling

10. Demolition

11. Upkeep

12. Maintenance – Painting and cleaning work

13. Drainage

ANNEX II

NON-EXHAUSTIVE LIST OF WORK INVOLVING PARTICULAR RISKS TO THE SAFETY AND HEALTH OF WORKERS REFERRED TO IN ARTICLE 3 (2), SECOND PARAGRAPH OF THE DIRECTIVE

1. Work which puts workers at risk of burial under earthfalls, engulfment in swampland or falling from a height, where the risk is particularly aggravated by the nature of the work or processes used or by the environment at the place of work or site.

2. Work which puts workers at risk from chemical or biological substances constituting a particular danger to the safety and health of workers or involving a legal requirement for health monitoring.

3. Work with ionizing radiation requiring the designation of controlled or supervised areas as defined in Article 20 of Directive 80/836/Euratom.

4. Work near high voltage power lines.

5. Work exposing workers to the risk of drowning.

6. Work on wells, underground earthworks and tunnels.

7. Work carried out by divers having a system of air supply.

8. Work carried out by workers in caisson with a compressed-air atmosphere.

9. Work involving the use of explosives.

10. Work involving the assembly or dismantling of heavy prefabricated components.

ANNEX III

CONTENT OF THE PRIOR NOTICE REFERRED TO IN ARTICLE 3 (3), FIRST PARAGRAPH OF THE DIRECTIVE I

1. Date of forwarding: ....

2. Exact address of the construction site: ....
    ....
    ....

3. Client(s) (name(s) and address(es)): ....
    ....
    ....

4. Type of project: ....

5. Project supervisor(s) (name(s) and address(es)): ....
    ....
    ....

6. Safety and health coordinators(s) during the project preparation stage (name(s) and address(es)): ....
    ....
    ....

7. Coordinator(s) for safety and health matters during the project execution stage (name(s) and address(es)): ....
    ....
    ....
    ....

8. Date planned for start of work on the construction site: ....
    ....

9. Planned duration of work on the construction site: ....
    ....

10. Estimated maximum number of workers on the construction site:
    ....

11. Planned number of contractors and self-employed persons on the construction site: ....
....

12. Details of contractors already chosen: ....
....
....
....
....
....
....
....

ANNEX IV (not included)

Appendix 2

HEALTH AND SAFETY (ENFORCING AUTHORITY) REGULATIONS 1989

SCHEDULE 1
MAIN ACTIVITIES WHICH DETERMINE WHETHER LOCAL
AUTHORITIES WILL BE ENFORCING AUTHORITIES

1. The sale or storage of goods for retail or wholesale distribution
    except –

    (a) where it is part of the business of a transport undertaking;

    (b) at container depots where the main activity is the storage of
        goods in the course of transport to or from dock premises, an
        airport or a railway;

    (c) where the main activity is the sale or storage for wholesale dis-
        tribution of any dangerous substance;

    (d) where the main activity is the sale or storage of water or sewage
        or their by-products or natural or town gas;

    and for the purposes of this paragraph where the main activity carried
    on in premises is the sale and fitting of motor car tyres, exhausts,
    windscreens or sunroofs the main activity shall be deemed to be the
    sale of goods.

2. The display or demonstration of goods at an exhibition for the pur-
    poses of offer or advertisement for sale.

3. Office activities.

4. Catering services.

5. The provision of permanent or temporary residential accommodation including the provision of a site for caravans or campers.

6. Consumer services provided in a shop except dry cleaning or radio and television repairs, and in this paragraph 'consumer services' means services of a type ordinarily supplied to persons who receive them otherwise than in the course of a trade, business or other undertaking carried on by them (whether for profit or not).

7. Cleaning (wet or dry) in coin operated units in launderettes and similar premises.

8. The use of a bath, sauna or solarium, massaging, hair transplanting, skin piercing, manicuring or other cosmetic services and therapeutic treatments, except where they are carried out under the supervision or control of a registered medical practitioner, a dentist registered under the Dentists Act 1984(a), a physiotherapist, an osteopath or a chiropractor.

9. The practice or presentation of the arts, sports, games, entertainment or other cultural or recreational activities except where carried on in a museum, art gallery or theatre or where the main activity is the exhibition of a cave to the public.

10. The hiring out of pleasure craft for use on inland waters.

11. The care, treatment, accommodation or exhibition of animals, birds or other creatures, except where the main activity is horse breeding or horse training at a stable, or is an agricultural activity or veterinary surgery.

12. The activities of an undertaker, except where the main activity is embalming or the making of coffins.

13. Church worship or religious meetings.

Appendix 3

NOTICE OF DECLARATION

(pursuant to regulation 4(4) of the Construction
(Design and Management) Regulations 1994)

To the Health and Safety Executive:

I, [name of declarant], [a director or the secretary or other duly authorised
officer] of [name of company] ('the Company' or 'the Firm'), HEREBY
DECLARE that

The [Company or Firm] shall act as the client until further notice for the
purposes of the Construction (Design and Management) Regulations 1994
in respect of the project known as . . . .
and located at . . . . (address).

The address for service of any documents on the client is [(address)].

. . . .
Signature

. . . .
Witness

Appendix 4

PARTICULARS TO BE NOTIFIED TO THE EXECUTIVE

1.   Date of forwarding.

2.   Exact address of the construction site.

3.   Name and address of the client or clients (see note).

4.   Type of project.

5.   Name and address of the planning supervisor.

6.   A declaration signed by or on behalf of the planning supervisor that he has been appointed.

7.   Name and address of the principal contractor.

8.   A declaration signed by or on behalf of the principal contractor that he has been appointed.

9.   Date planned for start of the construction phase.

10. Planned duration of the construction phase.

11. Estimated maximum number of people at work on the construction site.

12. Planned number of contractors on the construction site.

13. Name and address of any contractor or contractors already chosen.

Note: Where a declaration has been made in accordance with regulation 4(4), item 3 above refers to the client or clients on the basis that that declaration has not yet taken effect.

A SUGGESTED FORMAT FOR PROVIDING INFORMATION

PROJECT NOTICE [NO. ( )]
(served in accordance with regulation 7 of the
Construction (Design and Management) Regulations 1994)

ADDRESS OF THE SITE: . . . .
. . . .
. . . .
. . . .
. . . .

TYPE OF PROJECT: . . . .

THE PARTIES:
Name and address of the client, ...
or clients (or agent if a notice . . . .
has been served pursuant to . . . .
regulation 4(4)): . . . .

Name and address of the . . . .
planning supervisor: . . . .
. . . .
. . . .

Name and address of the . . . .
principal contractor: . . . .
. . . .
. . . .

Name and address of any . . . .
contractor or contractors . . . .
already selected: . . . .
. . . .
. . . .
. . . .
. . . .
. . . .

(continue on a separate sheet if necessary)

PLANNING DETAILS:
Date planned for start of the
construction phase: . . . .

Planned duration of the
construction phase: . . . .

Estimated maximum number of
persons at work on the
construction site: . . . .

Planned number of contractors
on the construction site: . . . .

DECLARATIONS:
I . . . . a duly authorised representative of the
planning supervisor named in this Notice do hereby declare that the
planning supervisor has been appointed pursuant to regulation 6(1)(a) for
the project described by this Notice.

Signed . . . . and dated . . . .

I . . . . a duly authorised representative of the
principal contractor named in this Notice do hereby declare that the
principal contractor has been appointed pursuant to regulation 6(1)(b) for
the project described by this Notice.

Signed . . . . and dated . . . .

Date of forwarding . . . .

Appendix 5

MANAGEMENT OF HEALTH AND SAFETY AT WORK REGULATIONS 1999

THE APPROVED CODE OF PRACTICE

Principles of prevention to be applied

Where an employer implements any preventive and protective measures he shall do so on the basis of the principles specified in Schedule 1 to these Regulations.

29 Employers and the self-employed need to introduce preventive and protective measures to control the risks identified by the risk assessment in order to comply with the relevant legislation. A set of principles to be followed in identifying the appropriate measures are set out in Schedule 1 to the Regulations and are described below. Employers and the self-employed should use these to direct their approach to identifying and implementing the necessary measures.

Guidance

30 In deciding which preventive and protective measures to take, employers and self-employed people should apply the following principles of prevention:

(a) if possible avoid a risk altogether, eg do the work in a different way, taking care not to introduce new hazards;

(b) evaluate risks that cannot be avoided by carrying out a risk assessment;

(c) combat risks at source, rather than taking palliative measures. So, if the steps are slippery, treating or replacing them is better than displaying a warning sign;

(d) adapt work to the requirements of the individual (consulting those who will be affected when designing workplaces, selecting work and personal protective equipment and drawing up working and safety procedures and methods of production). Aim to alleviate monotonous work and paced working at a predetermined rate, and increase the control individuals have over work they are responsible for;

(e) take advantage of technological and technical progress, which often offers opportunities for improving working methods and making them safer;

(f) implement risk prevention measures to form part of a coherent policy and approach. This will progressively reduce those risks that cannot be prevented or avoided altogether, and will take account of the way work is organised, the working conditions, the environment and any relevant social factors. Health and safety policy statements required under section 2(3) of the HSW Act should be prepared and applied by reference to these principles;

(g) give priority to those measures which protect the whole workplace and everyone who works there, and so give the greatest benefit (ie give collective protective measures priority over individual measures);

(h) ensure that workers, whether employees or self-employed, understand what they must do;

(i) the existence of a positive health and safety culture should exist within an organisation. That means the avoidance, prevention and reduction of risks at work must be accepted as part of the organisation's approach and attitude to all its activities. It should be recognised at all levels of the organisation, from junior to senior management.

31 These are general principles rather than individual prescriptive requirements. They should, however, be applied wherever it is reasonable to do so. Experience suggests that, in the majority of cases, adopting good practice will be enough to ensure risks are reduced sufficiently. Authoritative sources of good practice are prescriptive legislation, Approved Codes of Practice and guidance produced by

Government and HSE inspectors. Other sources include standards produced by standard-making organisations and guidance agreed by a body representing an industrial or occupational sector, provided the guidance has gained general acceptance. Where established industry practices result in high levels of health and safety, risk assessment should not be used to justify reducing current control measures.

Appendix 6

SUGGESTED CHECKLIST FOR THE HEALTH AND SAFETY PLAN

A non-exhaustive list of risks/matters that should be considered when preparing an assessment based on the Construction (Health, Safety and Welfare) Regulations 1996.

(a)   Falls.

(b)   Falling through fragile material.

(c)   Falling objects.

(d)   Stability of structures.

(e)   Demolition or dismantling.

(f)   Explosives.

(g)   Excavations.

(h)   Cofferdams and caissons.

(i)   Prevention of drowning.

(j)   Traffic routes.

(k)   Doors and gates.

(l)   Vehicles.

(m) Prevention of risk from fire.

(n)   Emergency routes and exits.

(o)   Welfare arrangements.

Appendix 7

THE REGULATIONS INCORPORATING THE AMENDMENTS INTRODUCED BY THE CONSTRUCTION (DESIGN AND MANAGEMENT) (AMENDMENT) REGULATIONS 2000 WHICH CAME INTO FORCE ON 2 OCTOBER 2000

THE CONSTRUCTION (DESIGN AND MANAGEMENT) REGULATIONS 1994 S.I. 1994/3140

Whereas the Health and Safety Commission has submitted to the Secretary of State, under section 11(2)(d) of the Health and Safety at Work etc. Act 1974[1] ("the 1974 Act"), proposals for the purpose of making regulations after the carrying out by the said Commission of consultations in accordance with section 50(3) of the 1974 Act;

And whereas the Secretary of State has made modifications to the said proposals under section 50(1) of the 1974 Act and has consulted the said Commission thereon in accordance with section 50(2) of that Act;

Now therefore, the Secretary of State, in exercise of the powers conferred on him by sections 15(1), (2), (3)(a) and (c), (4)(a), (6)(b) and (9), and 82(3)(a) of, and paragraphs 1(1)(c), 6(1), 14, 15(1), 20 and 21 of Schedule 3 to, the 1974 Act, and of all other powers enabling him in that behalf and for the purpose of giving effect to the said proposals of the said Commission with modifications as aforesaid, hereby makes the following Regulations:

Citation and commencement

1. These Regulations may be cited as the Construction (Design and Management) Regulations 1994 and shall come into force on 31st March 1995.

Interpretation

2. – (1) In these Regulations, unless the context otherwise requires –

"agent" in relation to any client means any person who acts as agent for a client in connection with the carrying on by the person of a trade, business or other undertaking (whether for profit or not);

"cleaning work" means the cleaning of any window or any transparent or translucent wall, ceiling or roof in or on a structure where such cleaning involves a risk of a person falling more than 2 metres;

"client" means any person for whom a project is carried out, whether it is carried out by another person or is carried out in-house;

"construction phase" means the period of time starting when construction work in any project starts and ending when construction work in that project is completed;

"construction work" means the carrying out of any building, civil engineering or engineering construction work and includes any of the following –

(a) the construction, alteration, conversion, fitting out, commissioning, renovation, repair, upkeep, redecoration or other maintenance (including cleaning which involves the use of water or an abrasive at high pressure or the use of substances classified as corrosive or toxic for the purposes of regulation 5 of the Carriage of Dangerous Goods by Road and Rail (Classification, Packaging and Labelling) Regulations 1994[2]), de-commissioning, demolition or dismantling of a structure,

(b) the preparation for an intended structure, including site clearance, exploration, investigation (but not site survey) and excavation, and laying or installing the foundations of the structure,

(c) the assembly of prefabricated elements to form a structure or the disassembly of prefabricated elements which, immediately before such disassembly, formed a structure,

(d) the removal of a structure or part of a structure or of any product or waste resulting from demolition or dismantling of a structure or from disassembly of prefabricated elements which, immediately before such disassembly, formed a structure, and

(e)   the installation, commissioning, maintenance, repair or removal of mechanical, electrical, gas, compressed air, hydraulic, telecommunications, computer or similar services which are normally fixed within or to a structure,

but does not include the exploration for or extraction of mineral resources or activities preparatory thereto carried out at a place where such exploration or extraction is carried out;

"contractor" means any person who carries on a trade, business or other undertaking (whether for profit or not) in connection with which he –

(a)   undertakes to or does carry out or manage construction work,

(b)   arranges for any person at work under his control (including, where he is an employer, any employee of his) to carry out or manage construction work;

"design" in relation to any structure includes drawing, design details, specification and bill of quantities (including specification of articles or substances) in relation to the structure;

"designer" means any person who carries on a trade, business or other undertaking in connection with which he prepares a design;

"developer" shall be construed in accordance with regulation 5(1);

"domestic client" means a client for whom a project is carried out not being a project carried out in connection with the carrying on by the client of a trade, business or other undertaking (whether for profit or not);

"health and safety file" means a file, or other record in permanent form, containing the information required by virtue of regulation 14(d);

"health and safety plan" means the plan prepared by virtue of regulation 15;

"planning supervisor" means any person for the time being appointed under regulation 6(1)(a);

"principal contractor" means any person for the time being appointed under regulation 6(1)(b);

"project" means a project which includes or is intended to include construction work;

"structure" means –

    (a)   any building, steel or reinforced concrete structure (not being a building), railway line or siding, tramway line, dock, harbour, inland navigation, tunnel, shaft, bridge, viaduct, waterworks, reservoir, pipe or pipe-line (whatever, in either case, it contains or is intended to contain), cable, aqueduct, sewer, sewage works, gasholder, road, airfield, sea defence works, river works, drainage works, earthworks, lagoon, dam, wall, caisson, mast, tower, pylon, underground tank, earth retaining structure, or structure designed to preserve or alter any natural feature, and any other structure similar to the foregoing, or

    (b)   any formwork, falsework, scaffold or other structure designed or used to provide support or means of access during construction work, or

    (c)   any fixed plant in respect of work which is installation, commissioning, de-commissioning or dismantling and where any such work involves a risk of a person falling more than 2 metres.

(2)   In determining whether any person arranges for a person (in this paragraph called "the relevant person") to prepare a design or to carry out or manage construction work regard shall be had to the following, namely –

    (a)   a person does arrange for the relevant person to do a thing where –

        (i)   he specifies in or in connection with any arrangement with a third person that the relevant person shall do that thing (whether by nominating the relevant person as a subcontractor to the third person or otherwise), or

(ii) being an employer, it is done by any of his employees in-house;

(b) a person does not arrange for the relevant person to do a thing where –

(i) being a self-employed person, he does it himself or, being in partnership it is done by any of his partners; or

(ii) being an employer, it is done by any of his employees otherwise than in-house, or

(iii) being a firm carrying on its business anywhere in Great Britain whose principal place of business is in Scotland, it is done by any partner in the firm; or

(iv) having arranged for a third person to do the thing, he does not object to the third person arranging for it to be done by the relevant person,

and the expressions "arrange" and "arranges" shall be construed accordingly.

(3) For the purposes of these Regulations –

(a) a project is carried out in-house where an employer arranges for the project to be carried out by an employee of his who acts, or by a group of employees who act, in either case, in relation to such a project as a separate part of the undertaking of the employer distinct from the part for which the project is carried out; and

(b) construction work is carried out or managed in-house where an employer arranges for the construction work to be carried out or managed by an employee of his who acts or by a group of employees who act, in either case, in relation to such construction work as a separate part of the undertaking of the employer distinct from the part for which the construction work is carried out or managed; and

    (c)   a design is prepared in-house where an employer arranges for the design to be prepared by an employee of his who acts, or by a group of employees who act, in either case, in relation to such design as a separate part of the undertaking of the employer distinct from the part for which the design is prepared.

(3A) Any reference in these Regulations to a person preparing a design shall include a reference to his employee or other person at work under his control preparing it for him; but nothing in this paragraph shall be taken to effect the application of paragraph (2).

(4)   For the purposes of these Regulations, a project is notifiable if the construction phase –

    (a)   will be longer than 30 days; or

    (b)   will involve more than 500 person days of construction work,

and the expression "notifiable" shall be construed accordingly.

(5)   Any reference in these Regulations to a person being reasonably satisfied –

    (a)   as to another person's competence is a reference to that person being satisfied after the taking of such steps as it is reasonable for that person to take (including making reasonable enquiries or seeking advice where necessary) to satisfy himself as to such competence; and

    (b)   as to whether another person has allocated or will allocate adequate resources is a reference to that person being satisfied that after the taking of such steps as it is reasonable for that person to take (including making reasonable enquiries or seeking advice where necessary) –

        (i)   to ascertain what resources have been or are intended to be so allocated; and

        (ii)   to establish whether the resources so allocated or intended to be allocated are adequate.

(6) Any reference in these Regulations to –

     (a) a numbered regulation or Schedule is a reference to the regulation in or Schedule to these Regulations so numbered; and

     (b) a numbered paragraph is a reference to the paragraph so numbered in the regulation in which the reference appears.

Application of regulations

3. – (1) Subject to the following paragraphs of this regulation, these Regulations shall apply to and in relation to construction work.

     (2) Subject to paragraph (3), regulations 4 to 12 and 14 to 19 shall not apply to or in relation to construction work included in a project where the client has reasonable grounds for believing that –

          (a) the project is not notifiable; and

          (b) the largest number of persons at work at any one time carrying out construction work included in the project will be or, as the case may be, is less than 5.

     (3) These Regulations shall apply to and in relation to construction work which is the demolition or dismantling of a structure notwithstanding paragraph (2).

     (4) These Regulations shall not apply to or in relation to construction work in respect of which the local authority within the meaning of regulation 2(1) of the Health and Safety (Enforcing Authority) Regulations 1989[3] is the enforcing authority.

     (5) Regulation 14(b) shall not apply to projects in which no more than one designer is involved.

     (6) Regulation 16(1)(a) shall not apply to projects in which no more than one contractor is involved.

     (7) Where construction work is carried out or managed in-house or a design is prepared in-house, then, for the purposes of

paragraphs (5) and (6), each part of the undertaking of the employer shall be treated as a person and shall be counted as a designer or, as the case may be, contractor, accordingly.

(8) Except where regulation 5 applies, regulations 4, 6, 8 to 12 and 14 to 19 shall not apply to or in relation to construction work included or intended to be included in a project carried out for a domestic client.

Clients and agents of clients

4. – (1) A client may appoint an agent or another client to act as the only client in respect of a project and where such an appointment is made the provisions of paragraphs (2) to (5) shall apply.

(2) No client shall appoint any person as his agent under paragraph (1) unless the client is reasonably satisfied that the person he intends to appoint as his agent has the competence to perform the duties imposed on a client by these Regulations.

(3) Where the person appointed under paragraph (1) makes a declaration in accordance with paragraph (4), then, from the date of receipt of the declaration by the Executive, such requirements and prohibitions as are imposed by these Regulations upon a client shall apply to the person so appointed (so long as he remains as such) as if he were the only client in respect of that project.

(4) A declaration in accordance with this paragraph –

    (a) is a declaration in writing, signed by or on behalf of the person referred to in paragraph (3), to the effect that the client or agent who makes it will act as client for the purposes of these Regulations; and

    (b) shall include the name of the person by or on behalf of whom it is made, the address where documents may be served on that person and the address of the construction site; and

    (c) shall be sent to the Executive.

(5)   Where the Executive receives a declaration in accordance with paragraph (4), it shall give notice to the person by or on behalf of whom the declaration is made and the notice shall include the date the declaration was received by the Executive.

(6)   Where the person referred to in paragraph (3) does not make a declaration in accordance with paragraph (4), any requirement or prohibition imposed by these Regulations on a client shall also be imposed on him but only to the extent it relates to any matter within his authority.

Requirements on developer

5. –   (1)   This regulation applies where the project is carried out for a domestic client and the client enters into an arrangement with a person (in this regulation called "the developer") who carries on a trade, business or other undertaking (whether for profit or not) in connection with which –

(a)   land or an interest in land is granted or transferred to the client; and

(b)   the developer undertakes that construction work will be carried out on the land; and

(c)   following the construction work, the land will include premises which, as intended by the client, will be occupied as a residence.

(2)   Where this regulation applies, with effect from the time the client enters into the arrangement referred to in paragraph (1), the requirements of regulations 6 and 8 to 12 shall apply to the developer as if he were the client.

Appointments of planning supervisor and principal contractor

6. –   (1)   Subject to paragraph (6)(b), every client shall appoint –

(a)   a planning supervisor; and

(b)   a principal contractor,

in respect of each project.

(2) The client shall not appoint as principal contractor any person who is not a contractor.

(3) The planning supervisor shall be appointed as soon as is practicable after the client has such information about the project and the construction work involved in it as will enable him to comply with the requirements imposed on him by regulations 8(1) and 9(1).

(4) The principal contractor shall be appointed as soon as is practicable after the client has such information about the project and the construction work involved in it as will enable the client to comply with the requirements imposed on him by regulations 8(3) and 9(3) when making an arrangement with a contractor to manage construction work where such arrangement consists of the appointment of the principal contractor.

(5) The appointments mentioned in paragraph (1) shall be terminated, changed or renewed as necessary to ensure that those appointments remain filled at all times until the end of the construction phase.

(6) Paragraph (1) does not prevent –

    (a) the appointment of the same person as planning supervisor and as principal contractor provided that person is competent to carry out the functions under these Regulations of both appointments; or

    (b) the appointment of the client as planning supervisor or as principal contractor or as both, provided the client is competent to perform the relevant functions under these Regulations.

Notification of project

7. – (1) The planning supervisor shall ensure that notice of the project in respect of which he is appointed is given to the Executive in accordance with paragraphs (2) to (4) unless the planning supervisor has reasonable grounds for believing that the project is not notifiable.

(2) Any notice required by paragraph (1) shall be given in writing or in such other manner as the Executive may from time to time

approve in writing and shall contain the particulars specified in paragraph (3) or, where applicable, paragraph (4) and shall be given at the times specified in those paragraphs.

(3) Notice containing such of the particulars specified in Schedule 1 as are known or can reasonably be ascertained shall be given as soon as is practicable after the appointment of the planning supervisor.

(4) Where any particulars specified in Schedule 1 have not been notified under paragraph (3), notice of such particulars shall be given as soon as is practicable after the appointment of the principal contractor and, in any event, before the start of construction work.

(5) Where a project is carried out for a domestic client then, except where regulation 5 applies, every contractor shall ensure that notice of the project is given to the Executive in accordance with paragraph (6) unless the contractor has reasonable grounds for believing that the project is not notifiable.

(6) Any notice required by paragraph (5) shall –

(a) be in writing or such other manner as the Executive may from time to time approve in writing;

(b) contain such of the particulars specified in Schedule 1 as are relevant to the project; and

(c) be given before the contractor or any person at work under his control starts to carry out construction work.

Competence of planning supervisor, designers and contractors

8. – (1) No client shall appoint any person as planning supervisor in respect of a project unless the client is reasonably satisfied that the person he intends to appoint has the competence to perform the functions of planning supervisor under these Regulations in respect of that project.

(2) No person shall arrange for a designer to prepare a design unless he is reasonably satisfied that the designer has the competence to prepare that design.

(3)   No person shall arrange for a contractor to carry out or manage construction work unless he is reasonably satisfied that the contractor has the competence to carry out or, as the case may be, manage, that construction work.

(4)   Any reference in this regulation to a person having competence shall extend only to his competence –

    (a)   to perform any requirement; and

    (b)   to conduct his undertaking without contravening any prohibition,

imposed on him by or under any of the relevant statutory provisions.

Provision for health and safety

9. – (1)   No client shall appoint any person as planning supervisor in respect of a project unless the client is reasonably satisfied that the person he intends to appoint has allocated or, as appropriate, will allocate adequate resources to enable him to perform the functions of planning supervisor under these Regulations in respect of that project.

(2)   No person shall arrange for a designer to prepare a design unless he is reasonably satisfied that the designer has allocated or, as appropriate, will allocate adequate resources to enable the designer to comply with regulation 13.

(3)   No person shall arrange for a contractor to carry out or manage construction work unless he is reasonably satisfied that the contractor has allocated or, as appropriate, will allocate adequate resources to enable the contractor to comply with the requirements and prohibitions imposed on him by or under the relevant statutory provisions.

Start of construction phase

10.   Every client shall ensure, so far as is reasonably practicable, that the construction phase of any project does not start unless a health and safety plan complying with regulation 15(4) has been prepared in respect of that project.

Client to ensure information is available

11. – (1) Every client shall ensure that the planning supervisor for any project carried out for the client is provided (as soon as is reasonably practicable but in any event before the commencement of the work to which the information relates) with all information mentioned in paragraph (2) about the state or condition of any premises at or on which construction work included or intended to be included in the project is or is intended to be carried out.

(2) The information required to be provided by paragraph (1) is information which is relevant to the functions of the planning supervisor under these Regulations and which the client has or could ascertain by making enquiries which it is reasonable for a person in his position to make.

Client to ensure health and safety file is available for inspection

12. – (1) Every client shall take such steps as it is reasonable for a person in his position to take to ensure that the information in any health and safety file which has been delivered to him is kept available for inspection by any person who may need information in the file for the purpose of complying with the requirements and prohibitions imposed on him by or under the relevant statutory provisions.

(2) It shall be sufficient compliance with paragraph (1) by a client who disposes of his entire interest in the structure if he delivers the health and safety file for the structure to the person who acquires his interest in the structure and ensures such person is aware of the nature and purpose of the health and safety file.

Requirements on designer

13. – (1) Except where a design is prepared in-house, no employer shall cause or permit any employee of his to prepare for him, and no self-employed person shall prepare, a design in respect of any project unless he has taken reasonable steps to ensure that the client for that project is aware of the duties to which the client is subject by virtue of these Regulations and of any practical guidance issued from time to time by the Commission with respect to the requirements of these Regulations.

(2) Every designer shall –

    (a) ensure that any design he prepares and which he is aware will be used for the purposes of construction work includes among the design considerations adequate regard to the need –

        (i) to avoid foreseeable risks to the health and safety of any person at work carrying out construction work or cleaning work in or on the structure at any time, or of any person who may be affected by the work of such a person at work,

        (ii) to combat at source risks to the health and safety of any person at work carrying out construction work or cleaning work in or on the structure at any time, or of any person who may be affected by the work of such a person at work, and

        (iii) to give priority to measures which will protect all persons at work who may carry out construction work or cleaning work at any time and all persons who may be affected by the work of such persons at work over measures which only protect each person carrying out such work;

    (b) ensure that the design includes adequate information about any aspect of the project or structure or materials (including articles or substances) which might affect the health or safety of any person at work carrying out construction work or cleaning work in or on the structure at any time or of any person who may be affected by the work of such a person at work; and

    (c) co-operate with the planning supervisor and with any other designer who is preparing any design in connection with the same project or structure so far as is necessary to enable each of them to comply with the requirements and prohibitions placed on him in relation to the project by or under the relevant statutory provisions.

(3) Sub-paragraphs (a) and (b) of paragraph (2) shall require the design to include only the matters referred to therein to the extent that it is reasonable to expect the designer to address them at the time the design is prepared and to the extent that it is otherwise reasonably practicable to do so.

Requirements on planning supervisor

14. The planning supervisor appointed for any project shall –

    (a) ensure, so far as is reasonably practicable, that the design of any structure comprised in the project –

        (i) includes among the design considerations adequate regard to the needs specified in heads (i) to iii) of regulation 13(2)(a), and

        (ii) includes adequate information as specified in regulation 13(2)(b);

    (b) take such steps as it is reasonable for a person in his position to take to ensure co-operation between designers so far as is necessary to enable each designer to comply with the requirements placed on him by regulation 13;

    (c) be in a position to give adequate advice to –

        (i) any client and any contractor with a view to enabling each of them to comply with regulations 8(2) and 9(2), and to (ii) any client with a view to enabling him to comply with regulations 8(3), 9(3) and 10;

    (d) ensure that a health and safety file is prepared in respect of each structure comprised in the project containing –

        (i) information included with the design by virtue of regulation 13(2)(b), and

        (ii) any other information relating to the project which it is reasonably foreseeable will be necessary to ensure the health and safety of any person at work who is carrying out or will carry out construction work or cleaning work in or on the structure or of any person who may be affected by the work of such a person at work;

(e) review, amend or add to the health and safety file prepared by virtue of sub-paragraph (d) of this regulation as necessary to ensure that it contains the information mentioned in that sub-paragraph when it is delivered to the client in accordance with sub-paragraph (f) of this regulation; and

(f) ensure that, on the completion of construction work on each structure comprised in the project, the health and safety file in respect of that structure is delivered to the client.

Requirements relating to the health and safety plan

15. – (1) The planning supervisor appointed for any project shall ensure that a health and safety plan in respect of the project has been prepared no later than the time specified in paragraph (2) and contains the information specified in paragraph (3).

(2) The time when the health and safety plan is required by paragraph (1) to be prepared is such time as will enable the health and safety plan to be provided to any contractor before arrangements are made for the contractor to carry out or manage construction work.

(3) The information required by paragraph (1) to be contained in the health and safety plan is –

(a) a general description of the construction work comprised in the project;

(b) details of the time within which it is intended that the project, and any intermediate stages, will be completed;

(c) details of risks to the health or safety of any person carrying out the construction work so far as such risks are known to the planning supervisor or are reasonably foreseeable;

(d) any other information which the planning supervisor knows or could ascertain by making reasonable enquiries and which it would be necessary for any contractor to have if he wished to show –

(i)   that he has the competence on which any person is required to be reasonably satisfied by regulation 8, or

(ii)   that he has allocated or, as appropriate, will allocate, adequate resources on which any person is required to be reasonably satisfied by regulation 9;

(e) such information as the planning supervisor knows or could ascertain by making reasonable enquiries and which it is reasonable for the planning supervisor to expect the principal contractor to need in order for him to comply with the requirement imposed on him by paragraph (4); and

(f) such information as the planning supervisor knows or could ascertain by making reasonable enquiries and which it would be reasonable for any contractor to know in order to understand how he can comply with any requirements placed upon him in respect of welfare by or under the relevant statutory provisions.

(4) The principal contractor shall take such measures as it is reasonable for a person in his position to take to ensure that the health and safety plan contains until the end of the construction phase the following features:

(a) arrangements for the project (including, where necessary, for management of construction work and monitoring of compliance with the relevant statutory provisions) which will ensure, so far as is reasonably practicable, the health and safety of all persons at work carrying out the construction work and all persons who may be affected by the work of such persons at work, taking account of –

(i)   risks involved in the construction work,

(ii)   any activity specified in paragraph (5); and

(b) sufficient information about arrangements for the welfare of persons at work by virtue of the project to enable any contractor to understand how he can comply with any requirements placed upon him in respect of welfare by or under the relevant statutory provisions.

(5) An activity is an activity referred to in paragraph (4)(a)(ii) if –

(a) it is an activity of persons at work; and

(b) it is carried out in or on the premises where construction work is or will be carried out; and

(c) either –

(i) the activity may affect the health or safety of persons at work carrying out the construction work or persons who may be affected by the work of such persons at work, or

(ii) the health or safety of the persons at work carrying out the activity may be affected by the work of persons at work carrying out the construction work.

Requirements on and powers of principal contractor

16. – (1) The principal contractor appointed for any project shall –

(a) take reasonable steps to ensure co-operation between all contractors (whether they are sharing the construction site for the purposes of regulation 11 of the Management of Health and Safety at Work Regulations 1999[4] or otherwise) so far as is necessary to enable each of those contractors to comply with the requirements and prohibitions imposed on him by or under the relevant statutory provisions relating to the construction work;

(b) ensure, so far as is reasonably practicable, that every contractor, and every employee at work in connection with the project complies with any rules contained in the health and safety plan;

(c) take reasonable steps to ensure that only authorised persons are allowed into any premises or part of premises where construction work is being carried out;

(d)   ensure that the particulars required to be in any notice given under regulation 7 are displayed in a readable condition in a position where they can be read by any person at work on construction work in connection with the project; and

(e)   promptly provide the planning supervisor with any information which –

  (i)   is in the possession of the principal contractor or which he could ascertain by making reasonable enquiries of a contractor, and

  (ii) it is reasonable to believe the planning supervisor would include in the health and safety file in order to comply with the requirements imposed on him in respect thereof in regulation 14, and

  (iii) is not in the possession of the planning supervisor.

(2) The principal contractor may –

(a)   give reasonable directions to any contractor so far as is necessary to enable the principal contractor to comply with his duties under these Regulations;

(b)   include in the health and safety plan rules for the management of the construction work which are reasonably required for the purposes of health and safety.

(3) Any rules contained in the health and safety plan shall be in writing and shall be brought to the attention of persons who may be affected by them.

Information and training

17. – (1) The principal contractor appointed for any project shall ensure, so far as is reasonably practicable, that every contractor is provided with comprehensible information on the risks to the health or safety of that contractor or of any employees or other persons under the control of that contractor arising out of or in connection with the construction work.

(2) The principal contractor shall ensure, so far as is reasonably practicable, that every contractor who is an employer provides

any of his employees at work carrying out the construction work with –

(a) any information which the employer is required to provide to those employees in respect of that work by virtue of regulation 10 of the Management of Health and Safety at Work Regulations 1999; and

(b) any health and safety training which the employer is required to provide to those employees in respect of that work by virtue of regulation 13(2)(b) of the Management of Health and Safety at Work Regulations 1999.

Advice from, and views of, persons at work

18. The principal contractor shall –

(a) ensure that employees and self-employed persons at work on the construction work are able to discuss, and offer advice to him on, matters connected with the project which it can reasonably be foreseen will affect their health or safety; and

(b) ensure that there are arrangements for the co-ordination of the views of employees at work on construction work, or of their representatives, where necessary for reasons of health and safety having regard to the nature of the construction work and the size of the premises where the construction work is carried out.

Requirements and prohibitions on contractors

19. – (1) Every contractor shall, in relation to the project –

(a) co operate with the principal contractor so far as is necessary to enable each of them to comply with his duties under the relevant statutory provisions;

(b) so far as is reasonably practicable, promptly provide the principal contractor with any information (including any relevant part of any risk assessment in his possession or control made by virtue of the Management of Health and Safety at Work Regulations 1999) which might affect the health or safety of any person at work carrying out the construction work or of any person who may be affected by the work of such a person at work or which might justify a review of the health and safety plan;

(c) comply with any directions of the principal contractor given to him under regulation 16(2)(a);

(d) comply with any rules applicable to him in the health and safety plan;

(e) promptly provide the principal contractor with the information in relation to any death, injury, condition or dangerous occurrence which the contractor is required to notify or report by virtue of the Reporting of Injuries, Diseases and Dangerous Occurrences Regulations 1995[5]; and

(f) promptly provide the principal contractor with any information which –

    (i) is in the possession of the contractor or which he could ascertain by making reasonable enquiries of persons under his control, and

    (ii) it is reasonable to believe the principal contractor would provide to the planning supervisor in order to comply with the requirements imposed on the principal contractor in respect thereof by regulation 16(1)(e), and

    (iii) which is not in the possession of the principal contractor.

(2) No employer shall cause or permit any employee of his to work on construction work unless the employer has been provided with the information mentioned in paragraph (4).

(3) No self-employed person shall work on construction work unless he has been provided with the information mentioned in paragraph (4).

(4) The information referred to in paragraphs (2) and (3) is –

(a) the name of the planning supervisor for the project;

    (b)   the name of the principal contractor for the project; and

    (c)   the contents of the health and safety plan or such part of it as is relevant to the construction work which any such employee or, as the case may be, which the self-employed person, is to carry out.

(5) It shall be a defence in any proceedings for contravention of paragraph (2) or (3) for the employer or self-employed person to show that he made all reasonable enquiries and reasonably believed –

    (a)   that he had been provided with the information mentioned in paragraph (4); or

    (b)   that, by virtue of any provision in regulation 3, this regulation did not apply to the construction work.

Extension outside Great Britain

20.   These Regulations shall apply to any activity to which sections 1 to 59 and 80 to 82 of the Health and Safety at Work etc. Act 1974 apply by virtue of article 7 of the Health and Safety at Work etc. Act 1974 (Application outside Great Britain) Order 1989[6] other than the activities specified in sub-paragraphs (b), (c) and (d) of that article as they apply to any such activity in Great Britain.

Exclusion of civil liability

21.   Breach of a duty imposed by these Regulations, other than those imposed by regulation 10 and regulation 16(1)(c), shall not confer a right of action in any civil proceedings.

Enforcement

22.   Notwithstanding regulation 3 of the Health and Safety (Enforcing Authority) Regulations 1989[7], the enforcing authority for these Regulations shall be the Executive.

Transitional provisions

23.   Schedule 2 shall have effect with respect to projects which have started, but the construction phase of which has not ended, when these Regulations come into force.

Repeals, revocations and modifications

24. – (1) Subsections (6) and (7) of section 127 of the Factories Act 1961[8] are repealed.

(2) Regulations 5 and 6 of the Construction (General Provisions) Regulations 1961[9] are revoked.

(3) The Construction (Notice of Operations and Works) Order 1965[10] is revoked.

(4) For item (i) of paragraph 4(a) of Schedule 2 to the Health and Safety (Enforcing Authority) Regulations 1989, the following item shall be substituted –

> "(i) regulation 7(1) of the Construction (Design and Management) Regulations 1994 (S.I.1994/3140) (which requires projects which include or are intended to include construction work to be notified to the Executive) applies to the project which includes the work; or".

Signed by order of the Secretary of State.

Phillip Oppenheim

Parliamentary Under Secretary of State, Department of Employment.

19th December 1994

Notes:

[1] 1974 c. 37; sections 15 and 50 were amended by the Employment Protection Act 1975 (c. 71), Schedule 15, paragraphs 6 and 16 respectively.

[2] S.I.1994/669.

[3] S.I.1989/1903.

[4] S.I.1999/3242.

[5] S.I.1995/3163.

[6] S.I.1989/840.

[7] S.I.1989/1903.

[8] 1961 c. 34.

[9] S.I.1961/1580; to which there are amendments not relevant to these Regulations.

[10] S.I.1965/221.

Index